Mesoscopic Route to Time Travel

Prosenjit Singha Deo

Mesoscopic Route to Time Travel

Springer

Prosenjit Singha Deo
S. N. Bose Centre
Kolkata, India

ISBN 978-981-16-4464-1 ISBN 978-981-16-4465-8 (eBook)
https://doi.org/10.1007/978-981-16-4465-8

This Springer imprint is published by the registered company Springer Nature Singapore Pte Ltd.
The registered company address is: 152 Beach Road, #21-01/04 Gateway East, Singapore 189721, Singapore

Dedicated to my parents.

Preface

As the title suggests, the book is on mesoscopic physics that interfaces between quantum mechanics and condensed matter physics. The author joined a Ph.D. program in 1991 and was introduced to the subject by his supervisor, Prof. A. M. Jayannavar. Not much was known at that time. Since then, he has almost single-handedly involved himself in addressing some of the key issues in mesoscopic physics which may be simply stated as follows: How can we understand the transport and thermodynamic properties of a system without relying on the axioms of statistical mechanics that are of equal apriori probability and ergodicity? And how can we control the quantum fluctuations to find device applications? Since much clarity has been achieved and many new principles have been revealed by the author's work spanning about 30 odd years, it makes sense to present the matter in the form of a book in a pedagogic manner. The main purpose is that the reader does not have to go through a myriad of literature and papers to learn the subject and carry it forward. It should be of help to new researchers, graduate and undergraduate students or even an experienced researcher who may be able to make some contributions if introduced to the issues and problems.

At various stages, the author was involved in collaborations with Dr. U. Satpathi, Dr. S. Mukherjee, Dr. S. Bandopadhyay, K. Meena, etc. He is also indebted to his teachers, Prof. A. M. Jayannavar, Prof. F. M. Peeters, Prof. M. Manninen and Prof. B. K. Talukdar, to name a few. He would also like to extend his courtesy to Prof. M. V. Berry, Prof. P. Panigrahi, Prof. S. Dattagupta, Prof. M. Buttiker, Prof. P. A. Sreeram, Prof. B. C. Gupta, Prof. S. Bhattacharjee, Prof. T. P. Pareek and Prof. P. K. Ghosh as some of them have provided valuable support, while others, lively discussions. The author's personal life was greatly enriched by his wife, parents, a host of friends (mostly birders, to specially mention Shyam Ghate, Bhaskar Das, Sanjay Thakur and Sopan Sen) and of course the lively music of Salil Chowdhury and Mark Knopfler that helped him cope with the hardships and frustrations of research . The author has personally experienced that while trained scientists think similar and are more predictable, an intelligent comment by a layperson or even a line in a song requires him to think out of the box.

Kolkata, India

Prosenjit Singha Deo

Contents

About the Author

Prof. Prosenjit Singha Deo is working at the S.N Bose National Centre for Basic Sciences in Kolkata, India, since 1999. He has successfully guided several Ph.D. students leading to their Ph.D. in physics and also mentored a few postdocs. His works are mostly analytical and theoretical, but he has worked in close collaboration with experimentalists and co-authored many scientific papers with experimentalists. He has consistently worked on the problems elaborated in the book for over 25 years and more. He has also taught several courses in physics at the level of M.Sc. Before joining S. N. Bose Centre, he spent about 3 years in Europe as a postdoctoral fellow. He did his Ph.D. in physics in 1996 from the Institute of Physics, Bhubaneswar, India.

Abbreviations

1D	1 Dimension
2D	2 Dimension
3D	3 Dimension
AB	Aharonov-Bohm
AC	Alternating Current
AD or A-D	Argand Diagram
BC	Burgers Circuit
DOS	Density of States
Eq.	Equation
Eqs.	Equations
Fig.	Figure
FSR	Friedel Sum Rule
HC	Hermitian conjugate
LDOS	Local Density of States
LPDOS	Local Partial Density of States
LPT	Larmor Precession Time
PDOS	Partial Density of States
Q1D	Quasi-One Dimension
Sc.	Schrodinger
WDT	Wigner Delay Time

Chapter 1
Open Systems

Mesoscopic physics has emerged as a frontier research area in physics due to the advent of recent technologies that can fabricate samples whose dimensions can go up to nano-meters. And at the same time we have seen tremendous achievements in subjecting these systems to very low temperatures so that quantum effects become dominant.

However, mesoscopic physics by definition is much broader than nano-physics in a rather elegant manner. Normally, the behavior of a physical system is determined by the intrinsic length and energy scales of the system. But when the sample dimensions become comparable to these length scales, then we can expect new phenomenon and new physics and that is the regime of mesoscopic physics. So novelty is the focus of mesoscopic physics. Therefore, as per definition, one may observe mesoscopic phenomenon at large and very large samples. However, the nano-fabrication techniques coupled with low-temperature cryogenics has proved very useful to thrust the subject forward and address one of the key issues in science, that of the mystery of the quantum world and how it crosses over to the classical world. This is a completely unknown territory and as our basic understanding of this regime becomes better, we can hope for major technological achievements. The quantum systems has given us many technologies like microwave ovens to clinical X-rays. The classical systems too give us technologies of aviation, travel and construction. And now we have high hopes on mesoscopic systems.

Microscopic properties of atoms and molecules can be harnessed by taking an ensemble as they obey certain principles of statistical mechanics only as an ensemble. So when we are talking about the mesoscopic regime, we are looking forward to new physics and principles that operate in the regime of samples that have to be studied in isolation or with minimal coupling and degrees of freedom such that we are not in the regime of text book style statistical mechanics. They may be naturally occurring systems like single atoms and molecules or fabricated systems like quantum rings and dots. Such studies can have two-fold implications. They can thrust forward our endeavor of down-scaling devices based on new physics and they can lead to a better

P. Singha Deo, *Mesoscopic Route to Time Travel*,
https://doi.org/10.1007/978-981-16-4465-8_1

understanding of the intermediate regime of quantum and classical worlds. In the absence of proper formalism, normally, one resorts to numerical techniques which do not allow us to pin down the exact principles operating at these scales. Besides, the truth of quantum mechanics is only in the continuous regime and infinite-dimensional vector spaces wherein operators like energy, position and momentum are unbounded operators and casting them in the form of a finite-dimensional matrix for numerical purposes is not merited until a proper scrutiny is made. Lots of studies should be made with the continuum approach. In this book, we will all along use the continuum approach.

We believe that readers interested in this book has gone through some of the basics outlined in the book (Datta 1995). This book gives a good relation between Boltzmann transport and mesoscopic transport. However, mesoscopic transport is much more fundamental in nature compared to the view presented in this book and leads to many new concepts and phenomenon not covered there. So here, we would like to take the reader in some such directions in a pedagogic manner. Thermodynamic properties of mesoscopic systems has been particularly ignored in this book and so is the role of evanescent modes, that we believe leads to many new phenomenon and will attract lot of attention in the future. Like, in S. Datta, we will restrict to electronic properties only because phononic properties in mesoscopic systems are mostly done in analogy with electronic properties and has not yet reached a level of maturity to be presented here in our opinion. Of course, the role of phonons on electronic properties has been presented to some extent in S. Datta, but that is again an alternate view to what will be also discussed here.

Typical non-equilibrium transport observables include entities like resistance, capacitance, inductance, etc. Different observables other than resistance has not received much attention in literature. Now that we have been able to understand mesoscopic resistance via the Landauer-Buttiker formalism, one must try to extend the philosophy to study other observables as well as the mysteries of nature. An isolated atom or molecule or any system (large or small) described by a Hamiltonian, does not exhibit these properties. The analogue in classical systems would be friction, viscosity, etc. Hamiltonian systems are completely reversible, while equilibrium thermodynamics or non-equilibrium transport require irreversibility. We do observe them in abundance in our everyday life and our technologies depend crucially on them. Sometimes they are a boon and sometimes a curse. So we always have to smuggle in something to get a fit with experiments and what we smuggle in is often very ill-defined.

Mesoscopic physics should be seen as the true ambassador to lead us to a fundamental understanding of such effects that live at the interface of the classical and quantum worlds. They can not only enrich our basic science but can also lead to novel technological developments. For example, Landauer conductance can be defined for a single molecule or atom and can be even probed experimentally. In some of the ideal samples, its predictions has been tested to an accuracy of one part in a billion which is far more accurate than predictions we get from solving closed Hamiltonian systems. Only other experimental data with similar agreement with theory is the Lamb shift. Another important fact brought out by Landauer conductance formula is that, in mesoscopic systems, we cannot define material-specific quantities like resistivity, permeability, dielectric constant, etc. This is essentially because these material-specific parameters are determined by the intrinsic length scales of the material such as coherence length, inelastic mean free path, screening length, etc. In mesoscopic systems, they face competition from external length scales like sample dimension, curvature of surfaces, etc. But we can define global properties like resistance, capacitance, magnetization, etc., for mesoscopic samples.

In mesoscopic physics, it is much more transparent as to how irreversibility sets in and allows us to treat semi-irreversible systems. In the subsequent section, we present a particular way of seeing this coming from the basic postulates of quantum statistical mechanics as outlined in the first three pages of Chap. 8 in K. Huang, Statistical physics, 3rd edition. So, there is no additional inputs like the arguments presented by Landauer in his original paper or the arguments given in the book by S. Datta. So we believe, readers who are psychologically set back by the new concepts involved in mesoscopic physics will be able to make themselves more comfortable and proceed.

After this section, we will mostly focus on thermodynamic effects. In the first two chapters, we have presented two approaches to mesoscopic systems, open and closed. This is a general philosophy in physics wherein one can see the closed system as a special case of an open system or one can see the open system as a special case of the closed system. One encounters this dichotomy formally in several forms like canonical system versus grand canonical system in statistical mechanics, closed intervals versus open intervals in mathematics, conservative systems versus dissipative systems in classical physics, eigenvalue problem versus decoherence in quantum mechanics, etc. In mesoscopic physics, this comes in the form of systems coupled to reservoirs versus isolated systems. Normally, in physics, one thinks the closed systems can be exactly understood as everything is known, whereas the open systems are those in which a few degrees of freedom are averaged out or ignored and most physical systems are open. In mesoscopic systems, this debate has always been very intense because averaging out of some degrees of freedom is rather demanding physically and conceptually. Landauer followed by Buttiker introduced the idea that mesoscopic open systems are coupled to a special kind of electron reservoirs that grasp physical situations remarkably well. Besides, at the non-interacting level, there are new ingredients in mesoscopic physics like magnetization due to Aharonov-Bohm effect and even-odd effect called parity effect. So in the first two chapters, we have discussed these in details. In this book, we will adopt the view that open systems are more fundamental than closed systems because of our recent findings that will be highlighted in Chaps. 5 and 6. In the second chapter, we will discuss the closed systems corresponding to the open systems in Chap. 1. There we will also discuss a new phase of the electron wavefunction that was first reported in a closed system but has dramatic effects in open systems.

If one has to adopt the hardcore approach of discarding definitions that follow from the Hamiltonian and modify them by averaging or integrating, then the Larmor clock approach resulting in a hierarchy of density of states is suited to treat mesoscopic systems. In this approach, one can define local entities like local partial density of states, emissivity, injectivity, etc., that cannot be defined from the Hamiltonian. However, their meanings were not very clear in terms of experiments that can be designed and pose a rather challenging mathematical concept that in quantum mechanics there are local descriptions and hidden degrees of freedom in the sense that one can talk of particular electrons that are going from one particular initial condition to one particular final condition and one can even talk of a number at a local point inside the system which gives the local state of the electron at that point. And there has remained special doubts related to the fact that these numbers may (although conclusive proof was absent) become negative. So in third chapter, we have given a detailed account of the Larmor clock. This is the formalism that allows us to directly access global properties like capacitance, inductance, AC response, non-linear response, etc. Any other formalism that first requires us to calculate resistivity, dielectric constant, permeability, susceptibility and other such material-specific parameters is not what we see as suited for mesoscopic systems for which by definition they do not exist. Of course, one can interpret terms in a broader sense but that is beyond the focus of this book. Therefore, we will restrict to a regime where not much has been done and there is a huge scope from the technological point of view as well as for basic science.

Before we go to Chaps. 5 and 6 where we solve the pertinent problems associated with Larmor clock, in Chap. 4, we briefly outline exact mathematical analysis to treat realistic physical systems and in particular cases we present a verification that Larmor clock is physically and mathematically correct. We also show how the even-odd parity effect can be utilized to reveal this. But, of course, these are particular instances which are therefore generalized in Chaps. 5 and 6. Mathematically, Larmor clock is consistent with the complex plane and the topology of the complex manifold. Physically, the Larmor clock is related to the physical propagation of a non-dispersive wavepacket in time. Negative partial density of states and negative traversal times are possible in real systems around Fano resonances. While the results in Chap. 5 can have a lot of technological applications, Chap. 6 will show the possibility of time travel and that can greatly enhance our basic understanding of the universe we live in. Necessary experimental observations will be quoted when necessary and such observations make this possibility very bright.

1.1 Transport Current

Consider a typical one-dimensional (1D) scattering problem discussed in text books and illustrated in Fig. 1.1. The equation of motion is the 1D Schrodinger (Sc.) equation

$$-\frac{\hbar^2}{2m}\frac{d^2\psi}{dx^2} + \phi(x)\psi(x) = E\psi(x) \tag{1.1}$$

In regions I and III of Fig. 1.1 where potential $\phi(x) = 0$, solution of this differential equation gives

$$\psi_I(x) = ae^{ikx} + be^{-ikx}$$

$$\psi_{III} = ce^{ikx'} + de^{-ikx'} \tag{1.2}$$

where $x' = x - l$. It is generally believed that the correct axiomatic approach to Quantum Mechanics was developed by von Neumann about a century ago but text books in quantum mechanics do not follow it. As a result, the concepts found in text books often run into contradictions and a student often takes a lot of time to come to terms with them, unless in the meantime he gets demotivated and quits. Text books are often found to state that the momentum operator is Hermitian and therefore momentum states are well defined. As a result, one can solve this kind of scattering problem in 1D by saying that a particle once transmitted never turns back because there is nothing to throw it back or reflect it back. This means that for the 1D scattering problem depicted in Fig. 1.1, in the negative semi-axis one sets a = 1, b = r (reflection amplitude), and c = t (transmission amplitude), d = 0 on the positive semi-axis. This picture works well when generalized to three-dimension (3D) in certain cases like the physics of particle colliders. Essentially, it is taken that scattered states in 3D are of the form $e^{i\vec{k}\cdot\vec{r}}$ but ironically, that does not satisfy the 3D Sc. equation (as obviously the radial and angular parts do not separate) and the time-reversed solutions are also ignored for no reason. These mathematical problems are bypassed in a phenomenological way to lead to a concept of scattering cross section and optical theorem that are practically very relevant. It is not guaranteed that the same phenomenology will describe physical situations in 1D (how such 1D physical situations arise will be described in Chap. 4). In fact, they do not. There are mathematical inconsistencies and physically not relevant.

Consider the 1D plane waves in Eq. 1.2, which unlike in 3D are true asymptotic solutions of 1D Sc. equation. Observables in quantum mechanics is given by self adjoint operators and symmetric (often wrongly called Hermitian in confusion with finite-dimensional matrices) operators are not enough to qualify. This is especially true for momentum operators and some cases are described by Bonneau et al. (2001) including the case of semi-infinite leads. For an elaborate discussion, one may see the lectures on quantum mechanics by Frederic Schuller in YouTube. But let us see if the momentum operator $\hat{p}$ is at least symmetric (for the simple-minded condensed matter community, we will refer to as Hermitian) which is always the initial test for an operator to qualify to describe an observable. Let us consider an inner product of the form

$$p_{fg} = \langle f|\, \hat{p}\, |g\rangle = \int_{-\infty}^{\infty} f^* \hat{p} g dx = \int_{-\infty}^{\infty} f^* \frac{\hbar}{i}\frac{dg}{dx} dx$$

$$= f^* \frac{\hbar}{i} g|_{-\infty}^{\infty} - \int_{-\infty}^{\infty} \frac{df^*}{dx}\frac{\hbar}{i} g dx$$

The first term is set to zero and is discussed below. Therefore,

$$p_{fg} = \int_{-\infty}^{\infty} \frac{df^*}{dx}[\frac{\hbar}{i}]^* g dx$$

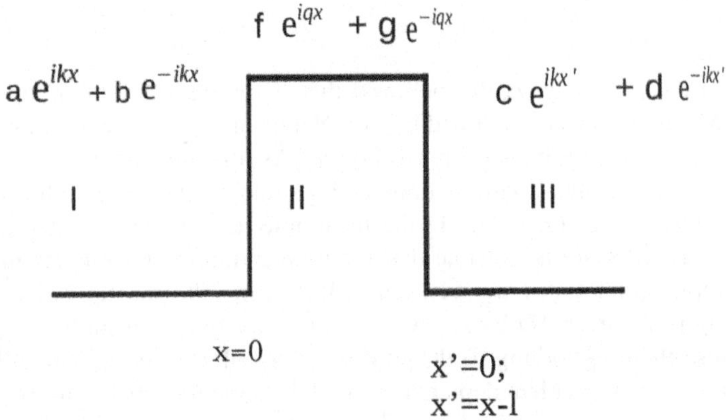

Fig. 1.1 A text book style scattering problem of a 1D rectangular barrier

Here, * means complex conjugation and $g^{**} = g$. Therefore,

$$p_{fg} = \int_{-\infty}^{\infty} [\frac{\hbar}{i}\frac{df}{dx}]^* g dx = p^*_{gf} \quad (1.3)$$

Let us come to the term set to zero, i.e., $f^* \frac{\hbar}{i} g|^{\infty}_{-\infty}$. If f and g are normalizable functions, then they go to zero at $\pm\infty$. But plane waves like the solutions of the Sc. equation in the leads (as in Eq. 1.2) are not normalizable. So at least for such plane waves in semi-infinite leads we cannot say that $\hat{p}$ is symmetric (or Hermitian). Besides most normalizable wavefunctions are of the form of a standing wave where symmetry does not guarantee a well-defined momentum state.

To quote a small verse from Upanishad where again generally people take the freedom to interpret and add words in between as they like. But "vishay prakashak gyanam" means "subject manifests in knowledge". If one is asked to find the area of a circle by fitting an infinite number of squares of arbitrary radius in it, a mathematician with all his mathematical rigor may insist that one should write down all the infinite terms in the series explicitly and then sum them up, a physicist like me would like to ask if all the terms in the series have a manifestation? If they manifest, then they do exist as a subject worth studying or as a source of knowledge and if they have no manifestation, then they do not exist, neither as a subject nor as a known entity or knowledge. So we can reduce the problem to that of a convergent series and solve it using a finite number of terms that actually manifests. Of course, in future, one may find manifestations of the ignored terms but for the problem in hand, if they do not manifest, then one can ignore them with respect to the given problem. The reader should take these words as an example and well-formulated mathematical problems must be taken very seriously with whatever mathematical rigor is required. We are not propounding subjectless manifestations in knowledge but rather saying that a well-formulated abstract mathematical problem can be a subject.

As stated before, it does not make any sense to use $a = 1$, $b = r$, $c = t$ and $d = 0$ which is how the problem is solved in text books. In condensed matter systems, we often do not have the freedom to rule out electrons incident simultaneously from the left as well as from the right. The leads may go to $\pm\infty$, but as the electron goes far, it reacts with the environment. Far away from the scatterer, a, b, c and d depend on the coordinates of the external world as well as on time (this time is obviously the classically recorded time at a point far from the scatterer where the electron starts interacting with the classical world). Thus, following Eqs. 8.5 and 8.6 in Kerson Huang,

$$\mid \bar{a} \mid^2 = 1, \mid \bar{d} \mid^2 = 1, \bar{ab} = \bar{cd} = \bar{a^*b} = \bar{ab^*} = \bar{cd^*} = \bar{c^*d} = \bar{d^*a} = \bar{a^*d} = 0. \tag{1.4}$$

So we will use these conditions and not $a = 1, b = r, c = t$ and $d = 0$. Basically, cross terms like $\lim_{x\to\pm\infty} \bar{f^*\frac{\hbar}{i}g} = 0.0$ implying symmetry. Thus, from Eq. 1.3, we can accept momentum is well defined in semi-infinite leads which will be further applied to practical systems in later chapters. Obviously, we are involving the environment and axioms of statistical mechanics, and what that means in terms of a mathematically well-defined self-adjoint momentum operator is left as an open question for future works at this point. Landauer's philosophy is essentially the same as that in Eq. 1.4 with a sharp demarcation between the classical world and quantum world which is very realistic and will also be described in details later. But if one is a little alienated by Landauer's arguments, then he can keep this text book picture in mind. However, the essential point is that one cannot explain Eq. 1.3 from a purely quantum mechanical viewpoint of Hamiltonian systems.

Close to the scatterer

$$\mid a \mid^2 = 1 \, and \, \mid d \mid^2 = 1 \tag{1.5}$$

as they have nothing to do with the scatterer, representing flux of incident particle equally probable from the left as well as the right. Thus, we are avoiding setting $d = 0$ by hand. a and d may have some initial phases which do not affect the analysis to follow. But close to the scatterer b and c will be determined by the scatterer. In Fig. 1.1, the amplitude b in region I can thus be written as

$$b = ar + dt'$$

at x = 0. This is because amplitude b at x = 0 contains the reflected part of a, i.e., ar and transmitted part of d, i.e., dt'. Similarly, one can write

$$c = dr' + at$$

at $x' = 0$. Here, r' and t' stand for scatterings due to incidence from the right. So total wavefunction in region III of Fig. 1.1 will be

$$\psi_{III} = dr'e^{ikx'} + ate^{ikx'} + de^{-ikx'}$$

Current density J in region III will be

$$J = \frac{e\hbar}{2im}(\psi^* \nabla \psi - \psi \nabla \psi^*)$$

$$= \frac{e\hbar}{m} Im(\psi^* \nabla \psi)$$

$$= \frac{e\hbar}{m} Im[(d^* r'^* e^{-ikx'} + a^* t^* e^{-ikx'} + d^* e^{ikx'}) \frac{d}{dx}(d r' e^{ikx'} + a t e^{ikx'} + d e^{-ikx'})]$$

Using Eqs. 1.4 and 1.5 and simplifying we get

$$= \frac{e\hbar}{m} Im[(ik)(\mid r' \mid^2 + \mid t \mid^2 - r'^* e^{-2ikx'} + r' e^{2ikx'} - 1)]$$

$$= \frac{e\hbar}{m} Im[(ik)(\mid r' \mid^2 + \mid t \mid^2 - \mid r' \mid e^{-2ikx' - i\eta} + \mid r' \mid e^{2ikx' + i\eta} - 1)]$$

$$= \frac{e\hbar}{m} Im[(ik)(\mid r' \mid^2 + \mid t \mid^2 + (2i) \mid r' \mid Sin(2kx' + \eta) - 1)]$$

where $r' = |r'|e^{i\eta}$. So, the current density on the right lead at T = 0 K for $e^{\pm ikx}$ incident wave (i.e., unit incident flux) is

$$J = \frac{e\hbar k}{m}[\mid r' \mid^2 + \mid t \mid^2 - 1] \tag{1.6}$$

Now current is of the form evn which implies the incident current in an energy interval dE is

$$dJ_{in} = ev\frac{dn}{dE}dE = ev\frac{1}{hv}dE \tag{1.7}$$

where $\frac{dn}{dE} = \frac{1}{hv}$ is the 1D density of states (DOS). And this results in the current through the sample to be (from Eq. 1.6)

$$dJ = \frac{e\hbar k}{m}(\mid r' \mid^2 + \mid t \mid^2 - 1)\frac{dn}{dE}dE$$

Basically, now instead of a unit incident flux, we have an incident flux of measure dn. Therefore

$$dJ = \frac{e\hbar k}{m}\frac{\mid r' \mid^2 + \mid t \mid^2 - 1}{hv}dE \tag{1.8}$$

$$= ev\frac{\mid r' \mid^2 + \mid t \mid^2 - 1}{hv}dE = \frac{e}{h}[\mid r' \mid^2 + \mid t \mid^2 - 1]dE \tag{1.9}$$

which was also derived by Landauer. Note that, comparing Eqs. 1.6 and 1.8 that incident current normalized by $\frac{1}{hv}$, is consistent with the 1D density of states. In this Eq. 1.8, $\frac{e\hbar k}{m}(\frac{-1}{hv})dE$ is the incident current before scattering (comparing with Eq. 1.7), while the current after scattering is $\frac{e\hbar k}{m}\frac{|r'|^2+|t|^2}{hv}dE$. dJ in Eq. 1.8 is therefore zero at zero temperature exhibiting Kirchhoff's law and hence the sign convention for the terms. A formal explanation of this sign convention can be seen from Eqs. 1.42 and 1.43 to be presented below. So the normalization constant for the wavefunctions will be

$$(\frac{1}{hv})^{1/2} = (\frac{2\pi m}{h^2 k})^{1/2} \tag{1.10}$$

which will automatically take care of the fact that the current at a given energy is carried by the available quantum states which equals to the DOS. This basic concept is essential to the understanding of mesoscopic transport and will be used subsequently in various forms. Now, at finite temperature (Datta 1995), a non-equilibrium current can be obtained by taking the temperature on the left different from that on the right modifying Eq. 1.9 to

$$dJ = \frac{e}{h}[f_r(E) \mid r' \mid^2 + f_l(E) \mid t \mid^2 - 1 f_r(E)] \tag{1.11}$$

Here, the first term is the reflection of electrons incident from the right and so will see the temperature of right reservoir and so on. Alternately, at zero temperature, one can take a chemical potential difference between the left and the right. If there is a current through the system, then the system can have a resistance which can be quantitatively calculated in the Landauer formalism and that as stated above is one of the greatest success of theoretical physics.

An incident wavefunction in a quantum mechanical scattering setup is rather correctly described by wavepackets and the plane waves in Eq. 1.2 are very correctly the Fourier components making the wavepacket. It is the wavepacket which is a true Hilbert space element. In the theory of quantum mechanics, the Fourier operator is more basic than the Sc. operator. Which means the Fourier components of a wavepacket in Hilbert space (a linear vector space) will always satisfy the Sc. equation. But not all solutions of the Sc. equation is admissible. A solution of the Sc. equation that cannot be a Fourier component making the wavepacket is a wrong solution. These solutions are often ignored in text books as "unphysical". In our opinion, the use of terms like physical versus unphysical may be misleading while discussing problems in quantum mechanics. Thus, Landauer's analysis is also theoretically correct for mesoscopic systems that are made of semiconductors like GaAs/AlGaAs in which electron propagation is that of free electrons with an effective mass.

Thermodynamics

Magnetization and polarization are two of the primary thermodynamic properties. The first is the response of a sample to magnetic field and the second is response to an applied electric field. Again they are emergent properties as we cannot define magnetization or polarization of an isolated atom or molecule. The general formalism to study such effects is discussed in Chap. 3. The relevance of that formalism can be seen in Mukherjee et al. (2011) where we microscopically derive and define quantum capacitance and numerically show that the formalism holds good in the presence of electron-electron interactions. In this book, we will mostly discuss magnetization as that leads to some dramatic results to be described in detail in Chaps. 5 and 6. Understanding magnetization in detail will help readers grasp the basic physics that might be extended to other unknown areas in the future. Mesoscopic response is predominantly dictated by interference effects like Aharonov-Bohm effect and Lorentz's forces play a negligible role. For example, mesoscopic disk can exhibit large magnetization due to a few Gauss, while a few Tesla of magnetic field is required to yield Hall effect-related responses. So we present below a very basic introduction to response of a 1D ring to a flux penetrating the ring, there being no magnetic field in the ring. This system is a natural choice because it can be treated analytically and the concepts can be clearly explained. For a quantum disk or any other structures, the same physics manifests and one can see the Aharonov-Bohm oscillations with a periodicity of Φ_0. But theoretically, such systems can only be solved numerically. There is a topological aspect to this problem which is usually not emphasized to beginners and will be explained below.

1.2 Aharonov-Bohm Effect

A charge particle in a magnetic field can exhibit this effect. Sc. equation for a particle (spinless) is

$$i\hbar\frac{\partial\psi}{\partial t} = H\psi \tag{1.12}$$

The Hamiltonian for the particle with charge q in magnetic field will be

$$H = \frac{1}{2m}(\vec{p} - \frac{q}{c}\vec{A})^2 + q\phi \tag{1.13}$$

$\vec{A}$ is the vector potential, ϕ is the scalar potential and $\vec{p} = \frac{\hbar}{i}\vec{\nabla}$. We know from Maxwell's equation that

$$\vec{B} = \vec{\nabla} \times \vec{A} \tag{1.14}$$

Gauge transformation is defined as

$$\vec{A}' \rightarrow \vec{A} + \vec{\nabla}\Lambda \tag{1.15}$$

$$\phi' \rightarrow \phi - \frac{1}{c}\frac{\partial\vec{A}}{\partial t} \tag{1.16}$$

$$\psi' \rightarrow U\psi \tag{1.17}$$

$$U = e^{\frac{iq\Lambda}{\hbar c}} \tag{1.18}$$

The last two are necessary for quantum particle only and not for classical particle. Let us try to understand the purpose of these transformations. The first of these in Eq. 1.15 is easy to see as $\vec{\nabla} X \vec{\nabla} \Lambda = 0$ for any scalar Λ, which means $\vec{A}$ and $\vec{A}'$ both give the same magnetic field $\vec{B}$. Schrödinger equation from Eqs. 1.12 and 1.13 after transformation is

$$i\hbar\frac{\partial\psi'}{\partial t} = [\frac{1}{2m}(\frac{\hbar}{i}\vec{\nabla} - \frac{q}{c}\vec{A}')^2 + q\phi']\psi' \tag{1.19}$$

Schrödinger equation from Eqs. 1.12 and 1.13 before transformation is

$$i\hbar\frac{\partial\psi}{\partial t} = [\frac{1}{2m}(\frac{\hbar}{i}\vec{\nabla} - \frac{q}{c}\vec{A})^2 + q\phi]\psi \tag{1.20}$$

The claim is that the potentials $\vec{A}$ and ϕ, as well as the Schrödinger equation are all equivalent before and after transformation. Which means we may work with either Eqs. 1.19 or 1.20, which ever is more convenient. Note that, for an electron, $q = -e$, while for a positron, $q = +e$. The equivalence of the two equations can be seen as follows. From Eq. 1.13,

$$H = \frac{1}{2m}(\vec{p}^2 - \frac{q}{c}\vec{p}\cdot\vec{A} - \frac{q}{c}\vec{A}\cdot\vec{p} + \frac{q^2}{c^2}\vec{A}^2) + q\phi \tag{1.21}$$

Classically, $\vec{p}$ and $\vec{A}$ commute. In quantum mechanics, since $\vec{p}$ is a differential operator,

$$\vec{p}\cdot\vec{A} = \frac{\hbar}{i}(\vec{\nabla}\cdot\vec{A}) + \vec{A}\cdot\vec{p} \tag{1.22}$$

The equivalence of LHS and RHS is with respect to their actions on a wavefunction. So

$$H = \frac{\vec{p}^2}{2m} + \frac{i\hbar q}{2mc}(\vec{\nabla}\cdot\vec{A}) - \frac{q}{mc}\vec{A}\cdot\vec{p} + \frac{q^2}{2mc^2}\vec{A}^2 + q\phi \tag{1.23}$$

Note that we may use Coulomb gauge by which $\vec{\nabla}\cdot\vec{A} = 0$ but that can be misleading when generalizing to rings with width and thickness. It is better to use the following identity:

$$(\frac{\hbar}{i}\vec{\nabla} - \frac{q}{c}\vec{A}')U\psi = U(\frac{\hbar}{i}\vec{\nabla} - \frac{q}{c}\vec{A})\psi \tag{1.24}$$

U is not a constant, but in the above identity, it behaves almost like a constant except $\vec{A}' \rightarrow \vec{A}$. The LHS is called gauge covariant derivative. Let us prove the identity stated in Eq. 1.24.

$$LHS = \frac{\hbar}{i}(\vec{\nabla}U)\psi + U\frac{\hbar}{i}\vec{\nabla}\psi - \frac{q}{c}\vec{A}U\psi - \frac{q}{c}(\vec{\nabla}\Lambda)U\psi$$

Note that we have used Eq. 1.15. Substituting for $U = e^{\frac{iq\Lambda}{\hbar c}}$ from Eq. 1.18

$$\frac{\hbar}{i}\frac{iq}{\hbar c}(\vec{\nabla}\Lambda)U\psi + U(\frac{\hbar}{i}\vec{\nabla} - \frac{q}{c}\vec{A})\psi - \frac{q}{c}(\vec{\nabla}\Lambda)U\psi = RHS$$

Using this identity, one can show that Eqs. 1.19 and 1.20 are the same as shown below. We start from Eq. 1.19, i.e.,

$$i\hbar\frac{\partial\psi'}{\partial t} = [\frac{1}{2m}(\frac{\hbar}{i}\vec{\nabla} - \frac{q}{c}\vec{A}')^2 + q\phi']\psi'$$

Using the identity in Eqs. 1.24 and 1.17,

$$i\hbar\frac{\partial}{\partial t}(U\psi) = \frac{1}{2m}(\frac{\hbar}{i}\vec{\nabla} - \frac{q}{c}\vec{A}')U(\frac{\hbar}{i}\vec{\nabla} - \frac{q}{c}\vec{A})\psi + q\phi' U\psi$$

or

$$i\hbar\frac{\partial}{\partial t}(U\psi) = \frac{1}{2m}(\frac{\hbar}{i}\vec{\nabla} - \frac{q}{c}\vec{A}')U\psi'' + q\phi' U\psi$$

where

$$(\frac{\hbar}{i}\vec{\nabla} - \frac{q}{c}\vec{A})\psi = \psi'' \tag{1.25}$$

or

$$i\hbar\frac{\partial}{\partial t}(U\psi) = \frac{1}{2m}U(\frac{\hbar}{i}\vec{\nabla} - \frac{q}{c}\vec{A})\psi'' + q\phi' U\psi$$

or

$$i\hbar\frac{\partial\psi}{\partial t} = \frac{1}{2m}(\frac{\hbar}{i}\vec{\nabla} - \frac{q}{c}\vec{A})\psi'' + q\phi'\psi$$

as U is independent of t. Thus using Eq. 1.25 and noting that in our case $\phi' = \phi$ (independent of t), we get

$$i\hbar\frac{\partial\psi}{\partial t} = \frac{1}{2m}(\frac{\hbar}{i}\vec{\nabla} - \frac{q}{c}\vec{A})^2\psi + q\phi\psi \tag{1.26}$$

Thus, we have shown that Eq. 1.19 is equivalent to Eq. 1.20 and we can solve any one of the two depending on convenience. It can be noted, that in Eq. 1.19, we have the freedom to set

$$\vec{A}' = \vec{A} + \vec{\nabla}\Lambda = 0 \tag{1.27}$$

as Λ can be chosen arbitrarily, which gives

$$i\hbar\frac{\partial\psi'}{\partial t} = [\frac{1}{2m}(\frac{\hbar}{i}\vec{\nabla})^2 + q\phi']\psi' \tag{1.28}$$

This is exactly the Sc. equation without the magnetic field but ψ' has an extra phase as can be seen from Eqs. 1.17 and 1.18 that accounts for the magnetic field in the center of the ring. Since there is no magnetic field in the ring, there is no Lorentz force acting on the electrons. Now the question is whether an extra phase in the complex wavefunction at all affect observables, specially when the probability is $|\psi'|^2$? The claim is that when closed trajectories are possible in the system then it will, and so this is a purely quantum interference phenomenon. This will result in an equilibrium current in the system that is not driven by any classical force. If the system does not allow any closed trajectories, then there will be no observable effect. There is no restriction on the nature of the closed trajectory. It could be circular, or square, or elliptical, or random walk that eventually comes back to the starting point. Of course, if we take a perfectly circular 1D ring then there is only the option to move circularly. But as soon as we want to generalize to a ring of finite thickness then one can choose from various types of such closed trajectories. So it is important to understand it from the topological point of view that we only need the possibility of trajectories that close on itself. Normally, to solve the Sc. equation in the presence of magnetic field, one does a gauge fixing. By gauge fixing, we make a particular choice of A_x, A_y and A_z, such that the contour integral of the phase picked up in going round the flux turns out to be definite and independent of the details of the path. There can be many different choices for the gauge fixing and none of the popular choices is appropriate for our problem. Let us demonstrate this for one such popular choice for gauge fixing called the Coulomb gauge, i.e., $\mathbf{\nabla} \cdot \mathbf{A} = 0$. Of course, we can write this down in polar form to simplify our calculations if our ring is perfectly circular and lying on a 2D plane. On the other hand, if we can restrict such that an electron has only the freedom to move in a square path of sides of length "a", then it may be more suitable to choose the Coulomb gauge in Cartesian coordinates. Coulomb gauge in Cartesian coordinates will be $\vec{\nabla}.\vec{A} = \partial_x A_x + \partial_y A_y + \partial_z A_z = 0$. That leads to

$$A_x = -B_0 y, \ A_y = 0, \ A_z = 0 \tag{1.29}$$

That in turn leads to a magnetic field

$$\vec{B} = \vec{\nabla} \times \vec{A} = \hat{k} B_0$$

$$B_z = \partial_x A_y - \partial_y A_x = B_0$$

$$B_x = \partial_y A_z - \partial_z A_y = 0$$

$$B_y = \partial_x A_z - \partial_z A_x = 0$$

Thus, Coulomb gauge in Cartesian coordinates give rise to the vector potential profile depicted in the Fig. 1.2. However, our choice of coordinate system should not determine the outcome of an observation, so one should be able to solve the situation of a perfectly circular ring with the kind of vector potential distribution shown in Fig. 1.2 and get the same result as that of gauge fixing in polar coordinates. So gauge

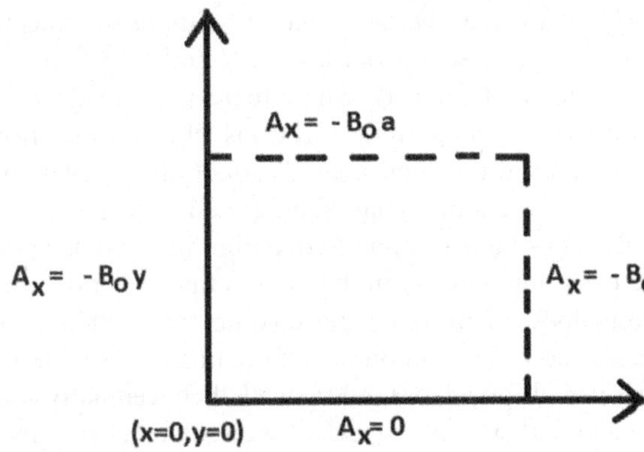

Fig. 1.2 Vector potential profile due to Coulomb gauge in Cartesian coordinates along a square path of sides of length a as given by Eq. 1.29

fixing may not be at all a good idea in case the ring has a finite thickness or for a disk because in such systems one can imagine all kinds of closed trajectories ranging from circular to square.

So one treats the problem using what is known in literature as twisted periodic boundary condition according to which a closed trajectory in a finite thickness ring means $\psi(x, y, z) = \psi(x + L_x, y, z)$. The analysis works for any arbitrary closed trajectory that one can imagine in a finite thickness ring (a disk also allows such closed trajectories as a ring is just a punctured disk). These trajectories are not directly related to electron motion in the system but effectively related through the concept of Feynman paths that will be further elaborated in Sect. 1.3. For the special case of a 1D ring, one can just drop the y and z degrees of freedom.

From Eqs. 1.17 and 1.18,

$$\psi'(x, y, z) = e^{-\frac{iq}{\hbar c}\Lambda}\psi(x, y, z) \tag{1.30}$$

$$\psi'(x + L_x, y, z) = e^{-\frac{iq}{\hbar c}\int_0^{L_x}\vec{\nabla}\Lambda\cdot\vec{dl}}\psi(x + L_x, y, z) = e^{-\frac{iq}{\hbar c}\int_0^{L_x}\vec{\nabla}\Lambda\cdot\vec{dl}}\psi(x, y, z) \tag{1.31}$$

because a closed trajectory implies $\psi(x + L_x, y, z) = \psi(x, y, z)$ for ψ defined in Eq. 1.20, while ψ' satisfies Eq. 1.28. Substituting $L_x = 0$ we get

$$\psi'(x, y, z) = e^{-\frac{iq}{\hbar c}\int_0^{0}\vec{\nabla}\Lambda\cdot\vec{dl}}\psi(x, y, z) = \psi(x, y, z) \tag{1.32}$$

Therefore, from Eqs. 1.31 and 1.32,

$$\psi'(x + L_x, y, z) = e^{-\frac{iq}{\hbar c}\int_0^{L_x}\vec{\nabla}\Lambda\cdot\vec{dl}}\psi'(x, y, z) \tag{1.33}$$

Using Eq. 1.27,

$$\psi'(x + L_x, y, z) = e^{\frac{iq}{\hbar c}\int_0^{L_x}\vec{A}\cdot\vec{dl}}\psi'(x, y, z) \tag{1.34}$$

or

$$\psi'(x + L_x, y, z) = e^{\frac{i2\pi\Phi}{\Phi_0}} \psi'(x, y, z) \tag{1.35}$$

$\Phi = \int_0^{L_x} \vec{A} \cdot dl$ is the flux through the ring and $\Phi_0 = \frac{hc}{e}$ is the flux quantum. Note that the only aspect of the ring geometry that has gone into this analysis is the fact that $x + L_x = x$ which means it can be an ellipse or any deformed version of a ring. To be mathematically correct, the analysis is for the topology of a ring and not ring geometry. In the rest of the chapter, we will use the simple system of a 1D ring because that can be solved analytically and the results can be elaborated with respect to the underlying physics. However, realistic systems like a ring of finite thickness or disk will also get magnetized by a fraction of a single flux quantum based on the same physical principle. For such systems, solving and demonstrating explicitly may involve some technical difficulties.

Now consider the system schematically shown in Fig. 1.3 that was first studied by Buttiker (1985). This is not an isolated ring but a ring connected to an electron reservoir making it a grand canonical system. The ring can exchange electrons with the reservoir and the way in which the reservoir is depicted is a very important aspect of mesoscopic physics. The situation considered in Sect. 1.1 corresponds to the case when there are two such electron reservoirs (not shown in the Fig. 1.1). One is to the left of region I and the other to the right of the region III in Fig. 1.1. These two reservoirs model the two battery terminals that drive the current through the system. Voltage probes (mesoscopic analogues of a voltmeter) and current probes (mesoscopic analogues of ammeters) also effectively act as additional electron reservoirs. Each has its own chemical potential and temperature and each can induce dephasing and decoherence (Datta 1995). It makes a lot of sense to study mesoscopic systems

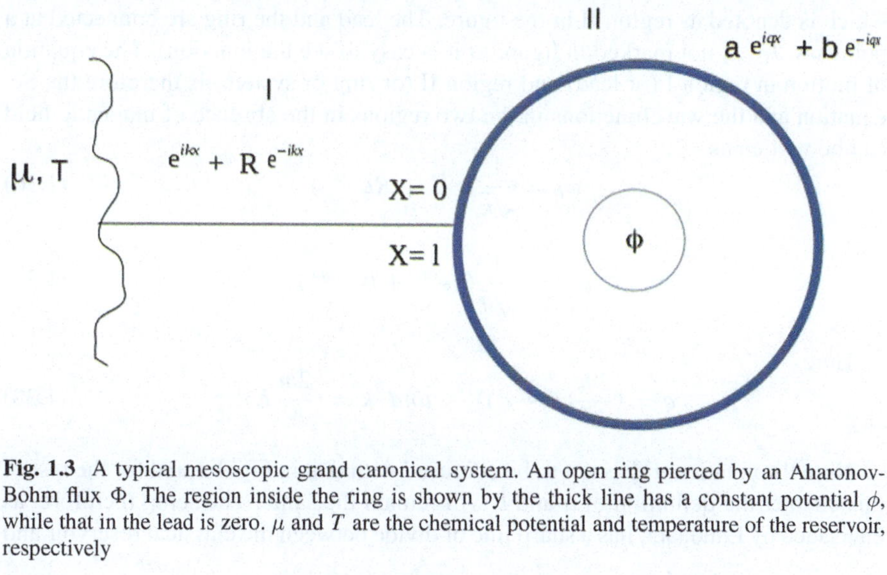

Fig. 1.3 A typical mesoscopic grand canonical system. An open ring pierced by an Aharonov-Bohm flux Φ. The region inside the ring is shown by the thick line has a constant potential ϕ, while that in the lead is zero. μ and T are the chemical potential and temperature of the reservoir, respectively

coupled to arbitrary number of reservoirs including one and two and all of them correspond to some physical situation that are experimentally relevant. With the help of Fig. 1.3 where there is only one reservoir, one may try to understand the novelty of mesoscopic physics. Single reservoir cannot drive the system out of equilibrium but can still introduce dephasing and decoherence in the system along with fluctuations in the particle number. The physical situation is that of a mesoscopic grand canonical system where the reservoir not only acts as a reservoir for electrons at a certain chemical potential and temperature (initially one may consider at zero temperature) but also naturally and inseparably serves as a source of decoherence due to the arguments of irreversibility given by Landauer. Now this is a well-established notion as can be seen in the book of S. Datta. The reservoir should be thought of as a classical system, while the system whose response we are trying to probe in this setup is the 1D ring pierced by an Aharonov-Bohm flux. The electrons inside the ring are purely described by quantum mechanics. If the ring is isolated then it will have some eigen functions and energy eigen states that can be obtained from Sc. equation and this will be presented in Chap. 2. These eigen states will carry a persistent current which is a purely equilibrium current as we will show. This current magnetizes the ring and has been observed in rings made of gold or semiconductors (Bluhm et al. 2009). The states in the ring of Fig. 1.3 will be affected by the reservoir that will give rise to dephasing and decoherence apart from fluctuations in particle number and thermal energy. Details of the results can be seen in the 1985 paper by Buttiker. Here, we will provide an exact and hence easy to understand quantum mechanical treatment of the problem as compared to the ad hoc theoretical treatment in the original paper by Buttiker. We will proceed as follows and which is known as Landauer-Buttiker scattering formalism. The reservoir injects electrons into the lead which is denoted as region I in the figure. It is just a 1D quantum wire where again the electrons are described by quantum mechanics. These electrons populate the states inside the ring which is denoted as region II in the figure. The lead and the ring are connected at a point say J_1 (J_1 not marked in figure as it is easy to see the junction). The equation of motion in region I (or lead) and region II (or ring or system) is therefore the Sc. equation and the wavefunctions in the two regions in the absence of magnetic field can be written as

$$\psi_I = \frac{1}{\sqrt{k}}(e^{ikx} + Re^{-ikx}) \tag{1.36}$$

$$\psi_{II} = \frac{1}{\sqrt{q}}(ae^{iqx} + be^{-iqx}) \tag{1.37}$$

Here

$$q = (\frac{2m}{\hbar^2}(E-\phi))^{1/2} \ and \ k = (\frac{2m}{\hbar^2}E)^{1/2} \tag{1.38}$$

Note the notation a, b is not to be confused with that in Fig. 1.1 and for the rest of this chapter the definition of a and b will remain the same. The setup therefore, as envisaged by Landauer, has a sharp line of divide between the classical reservoir and the quantum lead and this is now well understood to correspond to the reality when the total length of the lead and the ring is less than the inelastic scattering length of the electron. One can separately calculate the contact resistance for very short or long leads and sample resistance with different types of reservoirs and verify with experiments (Datta 1995). Consistency checks can be made with other formalisms like Kubo formula or Boltzmann transport or Keldysh formalism, etc. But one can bypass all the literature and research leading to it by adopting the simplified picture given in Sect. 1.1, that is one may remove the reservoir and assume the lead going to $-\infty$ with the condition that as the electron goes far from the ring it reacts with the environment. The normalization constants used in Eqs. 1.36 and 1.37 are consistent with Eq. 1.10. This therefore defines a pure quantum mechanical scattering problem and we outline below how we can calculate some observable quantities.

One can derive the current conserving boundary conditions starting from the time independent Sc. equation corresponding to Eq. 1.28 where $\phi' = \phi$ (as there is no time dependence), $q = 1$ and ψ' is written as ψ.

$$-\frac{\hbar^2}{2m}\nabla^2\psi(\vec{r}) + \phi(\vec{r})\psi(r) = E\psi(\vec{r}) \tag{1.39}$$

Integrating both sides over a volume V around the point where the lead meets the ring, we get

$$-\frac{\hbar^2}{2m}\int_V \nabla^2\psi(\vec{r})dV + \int_V \phi(\vec{r})\psi(\vec{r})dV = \int_V E\psi(\vec{r})dV \tag{1.40}$$

$$-\frac{\hbar^2}{2m}\int_V \vec{\nabla}.(\vec{\nabla}\psi(\vec{r}))dV + \int_V \phi(\vec{r})\psi(\vec{r})dV = \int_V E\psi(\vec{r})dV \tag{1.41}$$

Let S be the surface enclosing the volume V. In our 1D situation, this volume V will be therefore three 1D lines meeting at a point and the surface S will be the three end points of the figure shown in Fig. 1.4.

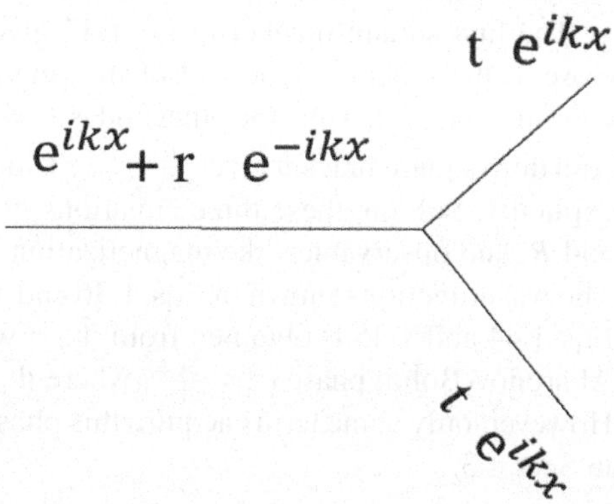

Fig. 1.4 The junction in Fig. 1.3 were the lead connects to the ring, in isolation

$$-\frac{\hbar^2}{2m}\lim_{V\to 0}\int_S(\vec{\nabla}\psi(\vec{r})).d\vec{S}+\lim_{V\to 0}\int_V\phi(\vec{r})\psi(\vec{r})dV=\lim_{V\to 0}\int_V E\psi(\vec{r})dV$$

Now as V is shrinking to 0, it follows that

$$\int_S(\vec{\nabla}\psi(\vec{r})).d\vec{S}=0$$

For the partial waves shown in Fig. 1.4, the integration becomes a sum and we can write,

$$\sum\vec{\nabla}\psi(\vec{r})=0 \tag{1.42}$$

This is Kirchhoff's law. Hence, in 1D, Eq. (1.42) can be written as

$$\sum_i\frac{d\psi_i}{dx_i}=0 \tag{1.43}$$

When this three prong structure of Fig. 1.4 is seen in isolation, then the wavefunctions obtained from Sc. equation is shown in Fig. 1.4. In that case, one can apply these boundary conditions to get $r=-1/3$ and $t=2/3$. While when it becomes a part of the system in Fig. 1.3, one has to use the wavefunctions depicted in Fig. 1.3. R will depend on r and t and the details can be found below. Single valuedness of wavefunction at the junction of Fig. 1.3 would imply

$$\frac{1}{\sqrt{k}}(1+R)=\frac{1}{\sqrt{q}}(a+be^{-i\alpha})=\frac{1}{\sqrt{q}}(ae^{iql+i\alpha}+be^{-iql}) \tag{1.44}$$

And Eq. 1.43 implies

$$[\frac{ik}{\sqrt{k}}(1-R)]+[\frac{iq}{\sqrt{q}}(-a+be^{-i\alpha})]+\frac{iq}{\sqrt{q}}[(ae^{iql+i\alpha}-be^{-iql})]=0 \tag{1.45}$$

The first square bracket in Eq. 1.45 gives $\frac{d\psi}{dx}|_{x=0^-}$ which is a point in the lead close to the junction. One end of the ring starting from the junction is assigned a coordinate $x=0$, while the other end is $x=l$. Second square bracket shows $\frac{d\psi}{dx}|_{x=0^+}$. And third square bracket gives $\frac{d\psi}{dx}|_{x=l^-}$. Thus, Eq. 1.45 is obtained by writing Eq. 1.43 explicitly. Solving these three equations, we can evaluate the three unknowns a, b and R, and observables like magnetization will depend on them as we show below. The wavefunctions shown in Eqs. 1.36 and 1.37 are in the absence of magnetic field. Eqs. 1.44 and 1.45 is obtained from these wavefunctions including the effect of the Aharonov-Bohm phase $\alpha=\frac{2\pi\Phi}{\Phi_0}$, where $\Phi_0=\frac{hc}{e}$ is the fundamental flux quantum. However, only some terms acquire this phase and others do not. We will explain this in Sect. 1.3.

Although the current in the ring in the absence of flux will be $|a|^2 - |b|^2$ and will obviously be zero, one cannot assume the same expression in the presence of flux. In the absence of flux a is the amplitude of clockwise moving electrons and b is the amplitude of anticlockwise electrons and they can only differ by a relative phase because none of them is preferred over the other. We give below a scheme to obtain the appropriate expression for current in the presence of flux, which can be extended to other complicated situations as well, for example, when we have evanescent modes inside the ring, or when some modes in the ring are propagating, while some are evanescent, etc. Such multichannel situations has been solved numerically as it can have technological applications (Mukherjee et al. 2020). The 1D system considered here can be solved analytically and is a good model to understand realistic systems. Current conservation implies that sum of currents arriving and leaving the junction is 0. So we quantum mechanically calculate the current $\frac{e\hbar}{2mi}[\psi^* \nabla \psi - HC]$ in region I and also at the two ends of region II. That for each region is written within square brackets and solved separately preserving current conservation in each step (for the time being assume the α dependence of the terms as in Eqs. 1.44 and 1.45 as their origin is explained in the next section).

$$\frac{1}{k}[(e^{ikx} + Re^{-ikx})^* \frac{d}{dx}(e^{ikx} + Re^{-ikx})]_{0^-} - \frac{1}{q}[(ae^{iqx} + be^{-iqx-i\alpha})^* \frac{d}{dx}(ae^{iqx} + be^{-iqx-i\alpha})]_{0^+} +$$

$$\frac{1}{q}[(ae^{iqx+i\alpha} + be^{-iqx})^* \frac{d}{dx}(ae^{iqx+i\alpha} + be^{-iqx})]_{l^-} - HC = 0 \quad (1.46)$$

Here, HC stands for Hermitian conjugate. Prefactors like $e\frac{\hbar}{2mi}$ can be added in the end. Clockwise current at $x = 0^+$ is leaving the junction and convention followed is currents leaving the junction are negative, while currents approaching the junction are positive according to the gradient convention in Eq. 1.42. The first bracketed term gives the current in the lead, that is immediate to the left of the junction in Fig. 1.3 or $x = 0^-$. The second bracketed term gives the current at a point in the ring, that is denoted as $x = 0^+$ in the figure. The third bracketed term gives the current at a point in the ring, that is denoted as $x = l^-$ in the figure. The equal signs follow from current conservation as there is no source in the junction. Simplifying we get

$$[\frac{ik}{k}(1 + R^*)(1 - R)] - [\frac{iq}{q}(a^* + b^* e^{i\alpha})(a - be^{-i\alpha})] +$$

$$[\frac{iq}{q}(a^* e^{-iql-i\alpha} + b^* e^{iql})(ae^{iql+i\alpha} - be^{-iql})] - HC = 0 \quad (1.47)$$

or

$$[i - iR + iR^* - i \mid R \mid^2] - [i \mid a \mid^2 - ia^* be^{-i\alpha} + ib^* ae^{i\alpha} - i \mid b \mid^2] +$$

$$[i \mid a \mid^2 - ia^*be^{-2iql-i\alpha} + ib^*ae^{2iql+i\alpha} - i \mid b \mid^2] - HC \tag{1.48}$$

The HC is written below

$$[-i + iR^* - iR + i \mid R \mid^2] - [-i \mid a \mid^2 + iab^*e^{i\alpha} - iba^*e^{-i\alpha} + i \mid b \mid^2] +$$
$$[-i \mid a \mid^2 + iab^*e^{2iql+i\alpha} - iba^*e^{-2iql-i\alpha} + i \mid b \mid^2] \tag{1.49}$$

Subtracting Eq. 1.49 from Eq. 1.48, we get 0 using the condition $\mid R \mid^2 = 1$. It is expected on physical grounds that the net current in the lead will be 0 and the current at $x = 0^+$ will be equal the current at $x = l^-$.

Let us consider only the second square bracket terms in Eqs. 1.48 and 1.49. That is

$$-i \mid a \mid^2 + ia^*be^{-i\alpha} - ib^*ae^{i\alpha} + i \mid b \mid^2$$

and

$$i \mid a \mid^2 - iab^*e^{i\alpha} + iba^*e^{-i\alpha} - i \mid b \mid^2$$

The difference between the two is

$$-2i \mid a \mid^2 + 2i \mid b \mid^2$$

which on including the $\frac{e\hbar}{2mi}$ gives me the current inside the ring to be

$$\frac{e\hbar}{m}(- \mid a \mid^2 + \mid b \mid^2) \tag{1.50}$$

This is always arbitrary to an overall sign for clockwise versus anticlockwise sign convention. So if $E > \phi$ then the current in the system or in the ring will be given by $\frac{e\hbar}{m}(|a|^2 - |b|^2)$. It can be easily calculated to check that if Φ is zero then there is no current in the system as is expected for an equilibrium situation. However, for $\Phi \neq 0$ the current is finite. It is an equilibrium current called persistent current, purely quantum mechanical in origin and further explanation of it will be given in Chap. 2. Here we would like to note that an interesting situation occurs when $E < \phi$ in Eq. 1.38. Then the persistent current is carried by evanescent modes. In this Situation, $q \to is$ and the expression for the current can be derived using the same scheme as before. Which means, we substitute $q \longrightarrow is$ and from Eqs. 1.36 and 1.37, ψ_I and ψ_{II} becomes

$$\psi_I = \frac{1}{\sqrt{k}}\left[e^{ikx} + Re^{-ikx}\right] \tag{1.51}$$

$$\psi_{II} = \frac{1}{\sqrt{(s)}}\left[ae^{i(is)x} + be^{-i(is)x}\right] \tag{1.52}$$

Applying boundary conditions of continuity and current conservation, (Eq. 1.43), analogues of Eqs. 1.44 and 1.45 become

$$\frac{1}{\sqrt{k}}(1+R)=\frac{1}{\sqrt{(s)}}(a+be^{-i\alpha})=\frac{1}{\sqrt{(s)}}(ae^{i(is)l+i\alpha}+be^{-i(is)l}) \qquad (1.53)$$

$$\left[\frac{1}{\sqrt{k}}ik(1-R)\right]+\left[\frac{1}{\sqrt{(s)}}i(is)(-a+be^{-i\alpha})\right]+\left[\frac{1}{\sqrt{(s)}}i(is)(ae^{i(is)l+i\alpha}-be^{-i(is)l})\right]=0 \qquad (1.54)$$

In Eq. 1.54, the first square bracket gives $\frac{d\psi_I}{dx}|_{x=0^-}$. The second square bracket shows $\frac{d\psi_{II}}{dx}|_{x=0^+}$. The third square bracket gives $\frac{d\psi_{II}}{dx}|_{x=l^-}$. So again we calculate for the wavefunctions in Eqs. 1.51 and 1.52 with the usual α dependence

$$\frac{e\hbar}{2mi}\left[\psi^*\nabla\psi-HC\right] \qquad (1.55)$$

and analogue of Eq. 1.46 becomes

$$\frac{1}{k}\left[(e^{ikx}+Re^{-ikx})^*\frac{d}{dx}(e^{ikx}+Re^{-ikx})\right]_{0^-}$$

$$-\frac{1}{s}\left[(ae^{i(is)x}+be^{-i(is)x-i\alpha})^*\frac{d}{dx}(ae^{i(is)x}+be^{-i(is)x-i\alpha})\right]_{0^+}$$

$$+\frac{1}{s}\left[(ae^{i(is)x+i\alpha}+be^{-i(is)x})^*\frac{d}{dx}(ae^{i(is)x+i\alpha}+be^{-i(is)x})\right]_{l^-}-HC=0 \qquad (1.56)$$

Evaluating the first three square brackets, we get

$$\left[i-iR+iR^*-i\mid R\mid^2\right]-\left[\mid a\mid^2-a^*be^{-i\alpha}+b^*ae^{i\alpha}-\mid b\mid^2\right]+$$

$$\left[\mid a\mid^2e^{-2sl}-a^*be^{-i\alpha}+b^*ae^{i\alpha}-\mid b\mid^2e^{2sl}\right] \qquad (1.57)$$

Considering the second square bracket term only for the current in the ring (the third square bracket can be alternately considered to give the same result) and substituting in the first term in Eq. 1.55, we get

$$-\frac{e\hbar}{2mi}\left[\mid a\mid^2-a^*be^{-i\alpha}+b^*ae^{i\alpha}-\mid b\mid^2\right] \qquad (1.58)$$

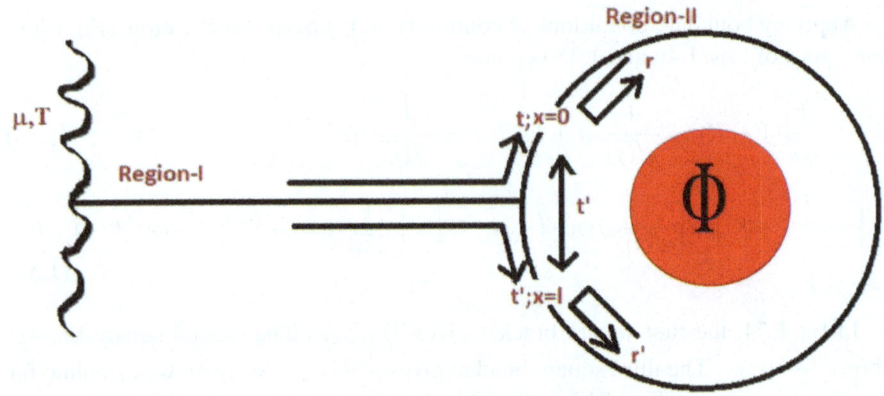

Fig. 1.5 Decomposition of the system in Fig. 1.3 into a three prong structure (depicted as region I) of Fig. 1.4 and a ring (depicted as region II). t and r are the transmission and reflection amplitudes of Fig. 1.4 which means $t = 2/3$ and $r = -1/3$. t' is related to t and r' is related to r in the manner explained in Eqs. 1.93 and 1.98. Usually, time reversal symmetry implies $t' = t$

Now subtracting the Hermitian conjugate, we get current expression in the ring

$$\frac{e\hbar}{mi}\left[ba^{*}e^{-i\alpha} - ab^{*}e^{i\alpha}\right] \tag{1.59}$$

1.3 Inclusion of Magnetic Field: Feynman Path Approach

In the Feynman path method, the full system is broken down into smaller parts as would be necessary and explained below. The system in Fig. 1.3 can be broken or decoupled into two parts. It is shown in Fig. 1.5 by breaking the lines at $x = 0^+$ and at $x = l^-$. Therefore the part depicted as region I is the three prong structure shown in Fig. 1.4. The part depicted as region II is the ring like structure enclosing the flux. Feynman path method provides a good connection between canonical solutions (leading to Eqs. 1.44 and 1.45) in quantum mechanics and physical motion (with a twist) resulting in currents through a sample. In the following, we present a calculation of the wavefunction inside the ring at $x = 0^+$ and $x = l^-$, using the Feynman path approach. This will also justify the way in which the AB phase enter the boundary conditions expressed in Eqs. 1.44 and 1.45. The twist is that one has to sum the amplitudes for all possible classical trajectories (weighted by their classical action) starting at the near end of the lead ($x = 0^-$) and ending at the point of interest where we intend to find the wavefunction ($x = 0^+$ or $x = l_-$) inside the ring. To systematically include all possible trajectories, we find a way to classify the trajectories and that can be done by the number of reflections involved and described in detail below as zero reflection terms, 1 reflection terms, etc.

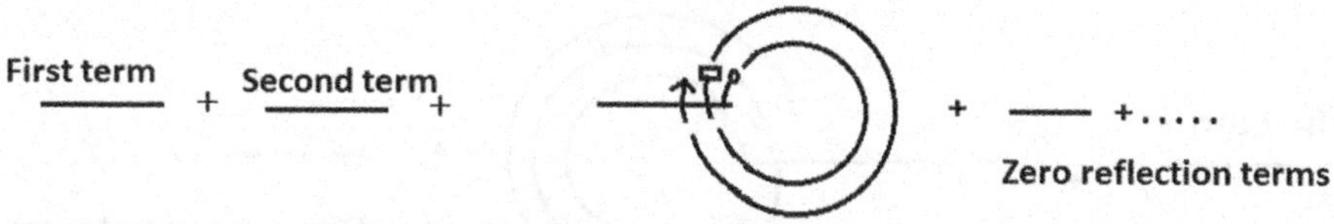

Fig. 1.6 The "First term" and the "Second term" are not drawn but can be figured out from the third term which is drawn explicitly. The third term correspond to a Feynman trajectory where an electron comes from the reservoir and at the junction it is transmitted up to the position of the small circle. It then makes a full circle along the ring to reach the position of the small rectangle. And then it continues to make another full circle in the ring to get to the position where the arrow head points. As per the coordinate choice depicted in Figs. 1.3 or 1.5 this position is $x = 0^+$. The first term will be the situation if this third term is truncated at the small circle. The second term will be the situation when this third term is truncated at the small rectangle. The first term will give the first term in the series leading to Eq. 1.60. The second term will give the second term in the series in Eq. 1.60. The third term will give the third term in the series in Eq. 1.60. For example, getting to the small circle will involve a transmission amplitude of $t = 2/3$ (first term in Eq. 1.60). And then make a full circle propagation with appropriate action require us to multiply a factor $e^{iql+i\alpha}$ and another factor of t' (which may or may not be the same as t as will be discussed subsequently) to get to the small rectangle (second term in Eq. 1.60). And finally to get to the point $x = 0^+$ at the end of the arrow require another factor of $e^{iql+i\alpha}t'$ (third term). None of the terms in Eq. 1.60 depend on r or r' and so all the terms are classified as zero reflection terms

Calculation of wavefunction at $x = 0^+$: Zero reflection terms depicted and explained in Fig. 1.6

$$\psi_0^{tt}(0^+) = t + te^{iql+i\alpha}t' + te^{iql+i\alpha}t'e^{iql+i\alpha}t' + \cdots$$
$$= \frac{t}{1 - t'e^{iql+i\alpha}} \tag{1.60}$$

Here, superscript tt stands for "through top", subscript 0 stands for "zero reflection term" and 0^+ in parenthesis denotes the coordinate where we intend to evaluate the wavefunction. These symbols will become more evident as we proceed. One reflection terms corresponding to Fig. 1.7

$$\psi_1^{tt}(0^+) = te^{iql+i\alpha}r'e^{iql-i\alpha} + te^{iql+i\alpha}r'e^{iql-i\alpha}t'e^{iql-i\alpha} + \cdots$$
$$= \frac{tr'e^{2iql}}{1 - t'e^{iql-i\alpha}} \tag{1.61}$$

Two reflection terms corresponding to Fig. 1.8

$$\psi_2^{tt}(0^+) = te^{iql+i\alpha}r'e^{iql-i\alpha}r + te^{iql+i\alpha}r'e^{iql-i\alpha}re^{iql+i\alpha}t' + \cdots$$
$$= \frac{trr'e^{2iql}}{1 - t'e^{iql+i\alpha}} \tag{1.62}$$

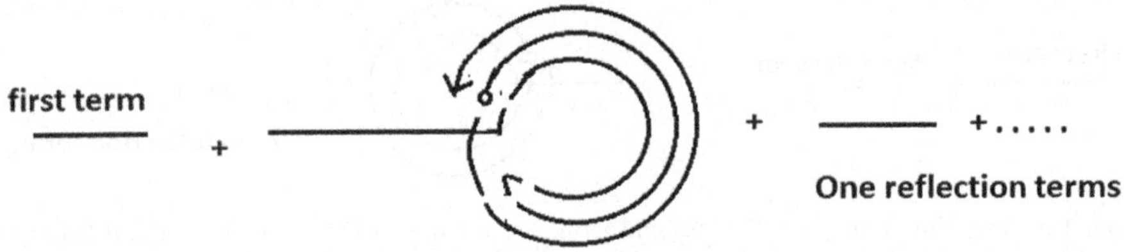

Fig. 1.7 One should be able to understand this figure by reading the caption of Fig. 1.6 and extend the logic. It depicts the Feynman paths leading to the series in Eq. 1.61

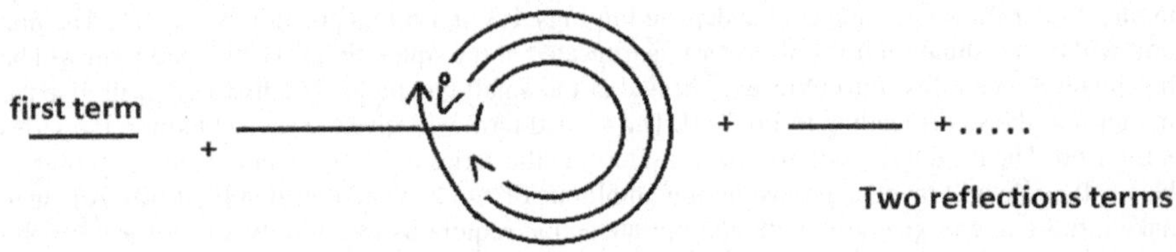

Fig. 1.8 It depicts the Feynman paths leading to the series in Eq. 1.62

Three reflection terms (self explanatory from earlier discussion)

$$\psi_3^{tt}(0^+) = te^{iql+i\alpha}r'e^{iql-i\alpha}re^{iql+i\alpha}r'e^{iql-i\alpha} + \cdots t'e^{iql-i\alpha} + \cdots$$

$$= \frac{tr'^2re^{4iql}}{1 - t'e^{iql-i\alpha}} \tag{1.63}$$

Four reflection terms (self explanatory from earlier discussion)

$$\psi_4^{tt}(0^+) = te^{iql+i\alpha}r'e^{iql-i\alpha}re^{iql+i\alpha}r'e^{iql-i\alpha}r + \cdots e^{iql+i\alpha}t' + \cdots$$

$$= \frac{tr^2r'^2e^{4iql}}{1 - t'e^{iql+i\alpha}} \tag{1.64}$$

Five reflection terms (self explanatory from earlier discussion)

$$\psi_5^{tt}(0^+) = te^{iql+i\alpha}r'e^{iql-i\alpha}re^{iql+i\alpha}r'e^{iql-i\alpha}re^{iql+i\alpha}r'e^{iql-i\alpha} + \cdots t'e^{iql-i\alpha} + \cdots$$

$$= \frac{tr'^3r^2e^{6iql}}{1 - t'e^{iql-i\alpha}} \tag{1.65}$$

Only five of the infinite number of terms are evaluated here to show that they form two geometric series. More details can be seen in the caption of Fig. 1.6. The even reflection terms in Eqs. 1.60, 1.62 and 1.64 form one series and the odd reflection terms in Eqs. 1.61, 1.63 and 1.65 form another.

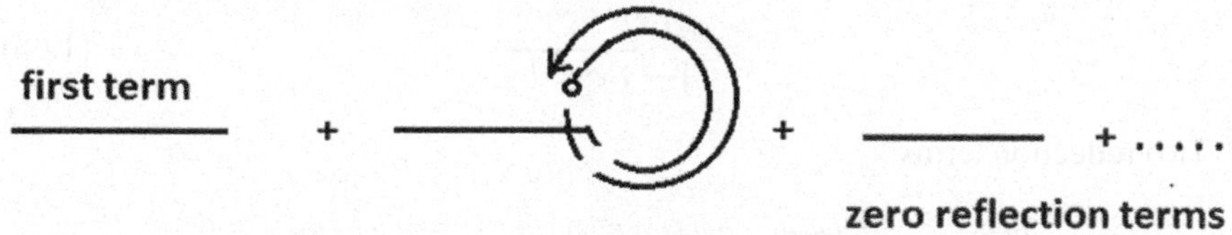

Fig. 1.9 It depicts the Feynman paths leading to the series in Eq. 1.67

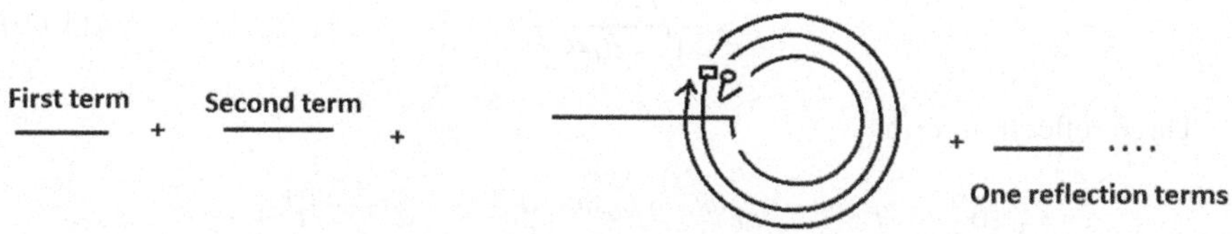

Fig. 1.10 It depicts the Feynman paths leading to the series in Eq. 1.68

So the contribution to the wavefunction inside the ring at the point $x = 0^+$ coming from these diagrams in Figs. 1.6, 1.7, 1.8, etc., will be the sum of the RHS of Eqs. 1.60 to 1.65

$$\psi^{tt}(0^+) = [\psi_0^{tt}(0^+) + \psi_2^{tt}(0^+) + \psi_4^{tt}(0^+) + \cdots] + [\psi_1^{tt}(0^+) + \psi_3^{tt}(0^+) + \psi_5^{tt}(0^+) + \cdots]$$

$$\psi^{tt}(0^+) = \frac{t}{1 - t'e^{iql+i\alpha}}(1 + rr'e^{2iql} + r^2r'^2e^{4iql} + \cdots).$$

$$+ \frac{tr'e^{2iql}}{1 - t'e^{iql-i\alpha}}(1 + r'r e^{2iql} + r^2r'^2e^{4iql} + \cdots)$$

or

$$\psi^{tt}(0^+) = \frac{t}{1 - t'e^{iql+i\alpha}}\frac{1}{1 - r'r e^{2iql}} + \frac{tr'e^{2iql}}{1 - t'e^{iql-i\alpha}}\frac{1}{1 - rr'e^{2iql}} \tag{1.66}$$

For the wavefunction at $x = 0^+$, one can get a different set of Feynman paths and the difference can be seen by comparing Fig. 1.6 with 1.9. These terms too can be classified by the number of reflections involved. Zero reflection terms (see Fig. 1.9)

$$\psi_0^{tb}(0^+) = t'e^{iql-i\alpha} + t'e^{iql-i\alpha}t'e^{iql-i\alpha} + \cdots$$

$$= \frac{t'e^{iql-i\alpha}}{1 - t'e^{iql-i\alpha}} \tag{1.67}$$

tb now stands for "through bottom". One reflection terms (see Fig. 1.10)

$$\psi_1^{tb}(0^+) = t'e^{iql-i\alpha}r + t'e^{iql-i\alpha}re^{iql+i\alpha}t' + \cdots$$

$$= \frac{t' r e^{iql-i\alpha}}{1 - t' e^{iql+i\alpha}} \tag{1.68}$$

Two reflection terms

$$\begin{aligned} \psi_2^{tb}(0^+) &= t' e^{iql-i\alpha} r e^{iql+i\alpha} r' e^{iql-i\alpha} + \cdots t' e^{iql-i\alpha} + \cdots \\ &= \frac{r' r t' e^{3iql-i\alpha}}{1 - t' e^{iql-i\alpha}} \end{aligned} \tag{1.69}$$

Three reflection terms

$$\begin{aligned} \psi_3^{tb}(0^+) &= t' e^{iql-i\alpha} r e^{iql+i\alpha} r' e^{iql-i\alpha} r + \cdots e^{iql+i\alpha} t' + \cdots \\ &= \frac{t' r' r^2 e^{3iql-i\alpha}}{1 - t' e^{iql+i\alpha}} \end{aligned} \tag{1.70}$$

Four reflection terms

$$\begin{aligned} \psi_4^{tb}(0^+) &= t' e^{iql-i\alpha} r e^{iql+i\alpha} r' e^{iql-i\alpha} r e^{iql+i\alpha} r' e^{iql-i\alpha} + \cdots t' e^{iql-i\alpha} + \cdots \\ &= \frac{t' r'^2 r^2 e^{5iql-i\alpha}}{1 - t' e^{iql-i\alpha}} \end{aligned} \tag{1.71}$$

Five reflection terms

$$\begin{aligned} \psi_5^{tb}(0^+) &= t' e^{iql-i\alpha} r e^{iql+i\alpha} r' e^{iql-i\alpha} r e^{iql+i\alpha} r' e^{iql-i\alpha} r + \cdots e^{iql+i\alpha} t' + \cdots \\ &= \frac{t' r^3 r'^2 e^{5iql-i\alpha}}{1 - t' e^{iql+i\alpha}} \end{aligned} \tag{1.72}$$

So the contribution to the wavefunction inside the ring at the point $x = 0^+$ coming from these diagrams will again form two geometric series,

$$\psi^{tb}(0^+) = [\psi_0^{tb}(0^+) + \psi_2^{tb}(0^+) + \psi_4^{tb}(0^+) + \cdots] + [\psi_1^{tb}(0^+) + \psi_3^{tb}(0^+) + \psi_5^{tb}(0^+) + \cdots$$

$$\psi^{tb}(0^+) = \frac{t' e^{iql-i\alpha}}{1 - t' e^{iql-i\alpha}} (1 + r' r e^{2iql} + r'^2 r^2 e^{4iql} + \cdots)$$

$$+ \frac{t' r e^{iql-i\alpha}}{1 - t' e^{iql+i\alpha}} (1 + r' r e^{2iql} + r^2 r'^2 e^{4iql} + \cdots)$$

$$\psi^{tb}(0^+) = \frac{t' e^{iql-i\alpha}}{1 - t' e^{iql-i\alpha}} \frac{1}{1 - r' r e^{2iql}} + \frac{t' r e^{iql-i\alpha}}{1 - t' e^{iql+i\alpha}} \frac{1}{1 - r' r e^{2iql}} \tag{1.73}$$

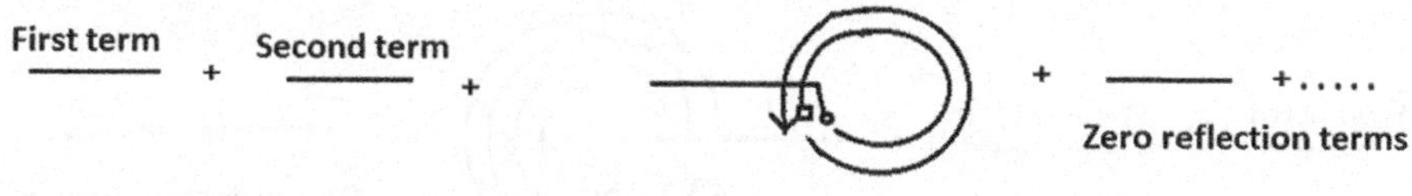

Fig. 1.11 It depicts the Feynman paths leading to the series in Eq. 1.77

The net wavefunction at $x = 0^+$

$$\psi(0^+) = \psi^{tt}(0^+) + \psi^{tb}(0^+)$$

$$= \frac{t}{1 - t'e^{iql+i\alpha}} \frac{1}{1 - rr'e^{2iql}} + \frac{tr'e^{2iql}}{1 - t'e^{iql-i\alpha}} \frac{1}{1 - rr'e^{2iql}}$$

$$+ \frac{t'e^{iql-i\alpha}}{1 - t'e^{iql-i\alpha}} \frac{1}{1 - rr'e^{2iql}} + \frac{t're^{iql-i\alpha}}{1 - t'e^{iql+i\alpha}} \frac{1}{1 - rr'e^{2iql}}$$

$$= \frac{t}{1 - rr'e^{2iql}}[\frac{1}{1 - t'e^{iql+i\alpha}} + \frac{r'e^{2iql}}{1 - t'e^{iql-i\alpha}}]$$

$$+ \frac{t'e^{iql-i\alpha}}{1 - rr'e^{2iql}}[\frac{1}{1 - t'e^{iql-i\alpha}} + \frac{r}{1 - t'e^{iql+i\alpha}}]$$

Thus, it is of the form $a + be^{-i\alpha}$ where

$$a = \frac{t}{1 - rr'e^{2iql}}[\frac{1}{1 - t'e^{iql+i\alpha}} + \frac{r'e^{2iql}}{1 - t'e^{iql-i\alpha}}] \tag{1.74}$$

$$b = \frac{t'e^{iql}}{1 - rr'e^{2iql}}[\frac{1}{1 - t'e^{iql-i\alpha}} + \frac{r}{1 - t'e^{iql+i\alpha}}] \tag{1.75}$$

This explains the second term in Eq. 1.44 and the second bracketed term in Eq. 1.45 where the normalization constant is added. Also note that a and b as obtained above is completely known as t, t', r and r' are known (see further Eq. 1.76 below). Thus, plugging them in Eq. 1.50, we can directly get the current in the ring without having to solve the simultaneous equations in 1.44 and 1.45. Also note that

$$r'e^{2iql} = r \tag{1.76}$$

as we will explain below implying that in the absence of flux there will be no current in the ring. And that also implies that the flux will break the symmetry between a and b to give rise to a current in the ring which will magnetize the ring. We would like to emphasize again that this is a purely equilibrium current due to quantum interference and Aharonov-Bohm effect. It does not require any classical force to sustain it and is a perpetual current. At this point, one may dive into the vast amount of literature

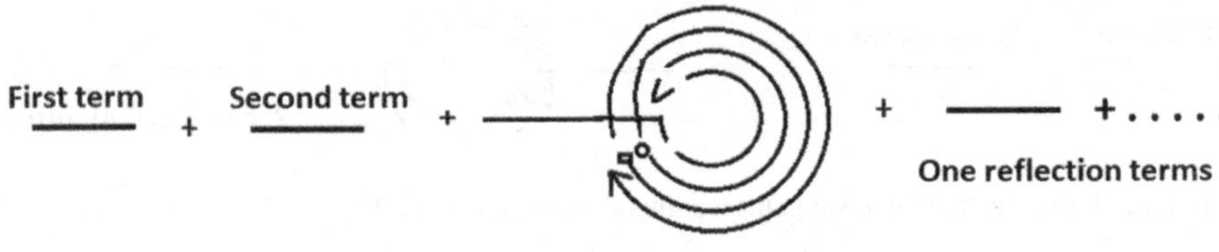

Fig. 1.12 It depicts the Feynman paths leading to the series in Eq. 1.78

to see how the current flows in the ring, but the heat is dissipated in the reservoir, implying a sharp spatial separation between the two processes, the first one being quantum and the second being classical. One may make the length of the lead tending to zero as that length never entered our analysis. Perpetual currents cannot survive in a dissipative ring. The length of the ring is less than the inelastic mean free path of the electron and these currents has been observed experimentally. However, strange it may seem, it is this picture of a reservoir that is physical and will be extended to real 3D systems in Chap. 4, while the view presented in Sect. 1.1 is an idealization to the situation of 1D semi-infinite leads. Now the literature is so extensive that we are not making any attempt to refer such a well-established phenomenon other than the book by S. Datta. The simple analysis presented here from first principles should make it clear.

Calculation of wavefunction at $x = l^-$: In the same way, we proceed to calculate the wavefunction at $x = l^-$. Zero reflection terms (see Fig. 1.11)

$$\psi_0^{tb}(l^-) = t' + t'e^{iql-i\alpha}t' + t'e^{iql-i\alpha}t'e^{iql-i\alpha}t' + \cdots$$

$$= \frac{t'}{1 - t'e^{iql-i\alpha}} \tag{1.77}$$

One reflection terms (see Fig. 1.12)

$$\psi_1^{tb}(l^-) = t'e^{iql-i\alpha}re^{iql+i\alpha} + t'e^{iql-i\alpha}re^{iql+i\alpha}t'e^{iql+i\alpha} + \cdots$$

$$= \frac{t're^{2iql}}{1 - t'e^{iql+i\alpha}} \tag{1.78}$$

Two reflection terms

$$\psi_2^{tb}(l^-) = t'e^{iql-i\alpha}re^{iql+i\alpha}r' + t'e^{iql-i\alpha}re^{iql+i\alpha}r'e^{iql-i\alpha}t' + \cdots$$

$$= \frac{t'rr'e^{2iql}}{1 - t'e^{iql-i\alpha}} \tag{1.79}$$

Three reflection terms

$$\psi_3^{tb}(l^-) = t'e^{iql-i\alpha}re^{iql+i\alpha}r'e^{iql-i\alpha}re^{iql+i\alpha} + \cdots t'e^{iql+i\alpha} + \cdots$$

$$= \frac{t'r^2r'e^{4iql}}{1 - t'e^{iql+i\alpha}} \tag{1.80}$$

Four reflection terms

$$\psi_4^{tb}(l^-) = t'e^{iql-i\alpha}re^{iql+i\alpha}r'e^{iql-i\alpha}re^{iql+i\alpha}r' + \cdots e^{iql-i\alpha}t' + \cdots$$

$$= \frac{t'r^2r'^2e^{4iql}}{1 - t'e^{iql-i\alpha}} \tag{1.81}$$

Five reflection terms

$$\psi_5^{tb}(l^-) = t'e^{iql-i\alpha}re^{iql+i\alpha}r'e^{iql-i\alpha}re^{iql+i\alpha}r'e^{iql-i\alpha}re^{iql+i\alpha}$$

$$+ \cdots t'e^{iql+i\alpha} + \cdots$$

$$= \frac{t'r^3r'^2e^{6iql}}{1 - t'e^{iql+i\alpha}} \tag{1.82}$$

So the contribution to the wavefunction inside the ring at the point $x = l^-$ coming from these diagrams will form two geometric series. One of them is given by the Eqs. 1.77, 1.79, 1.81 and so on and the other by the Eqs. 1.78, 1.80, 1.82 and so on to give

$$\psi^{tb}(l^-) = [\psi_0^{tb}(l^-) + \psi_2^{tb}(l^-) + \psi_4^{tb}(l^-) + \cdots] + [\psi_1^{tb}(l^-) + \psi_3^{tb}(l^-) + \psi_5^{tb}(l^-) + \cdots]$$

$$\psi^{tb}(l^-) = \frac{t'}{1 - t'e^{iql-i\alpha}}(1 + rr'e^{2iql} + r^2r'^2e^{4iql} + \cdots)$$

$$+ \frac{t're^{2iql}}{1 - t'e^{iql+i\alpha}}(1 + rr'e^{2iql} + r^2r'^2e^{4iql} + \cdots)$$

$$\psi^{tb}(l^-) = \frac{t'}{1 - t'e^{iql-i\alpha}}\frac{1}{1 - rr'e^{2iql}} + \frac{t're^{2iql}}{1 - t'e^{iql+i\alpha}}\frac{1}{1 - rr'e^{2iql}} \tag{1.83}$$

As in the previous case, there is another set of diagrams for the wavefunction at $x = l^-$. Zero reflection terms (see Fig. 1.13)

$$\psi_0^{tt}(l^-) = te^{iql+i\alpha} + te^{iql+i\alpha}t'e^{iql+i\alpha} + \cdots$$

$$= \frac{te^{iql+i\alpha}}{1 - t'e^{iql+i\alpha}} \tag{1.84}$$

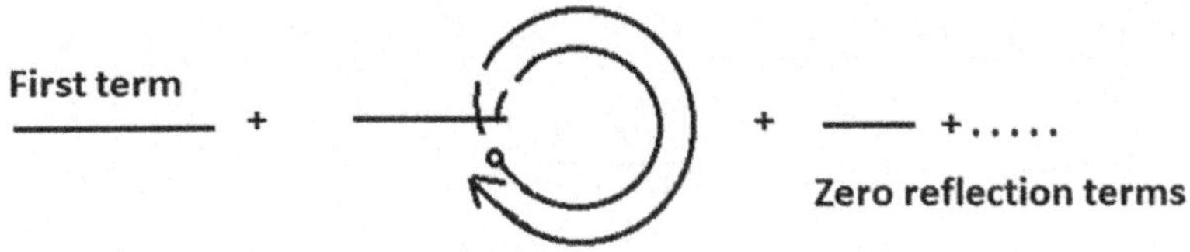

Fig. 1.13 It depicts the Feynman paths leading to the series in Eq. 1.84

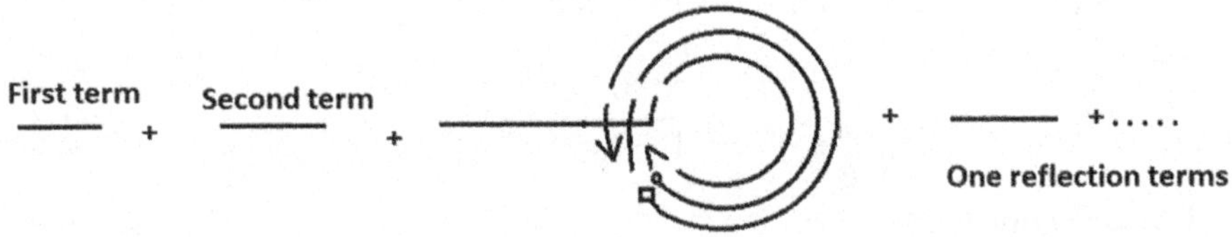

Fig. 1.14 It depicts the Feynman paths leading to the series in Eq. 1.85

One reflection terms (see Fig. 1.14)

$$\psi_1^{tt}(l^-) = te^{iql+i\alpha}r' + te^{iql+i\alpha}r'e^{iql-i\alpha}t' + \cdots$$
$$= \frac{tr'e^{iql+i\alpha}}{1 - t'e^{iql-i\alpha}} \tag{1.85}$$

Two reflection terms

$$\psi_2^{tt}(l^-) = te^{iql+i\alpha}r'e^{iql-i\alpha}re^{iql+i\alpha} + te^{iql+i\alpha}r'e^{iql-i\alpha}re^{iql+i\alpha}t'e^{iql+i\alpha} + \cdots$$
$$= \frac{tr're^{3iql+i\alpha}}{1 - t'e^{iql+i\alpha}} \tag{1.86}$$

Three reflection terms

$$\psi_3^{tt}(l^-) = te^{iql+i\alpha}r'e^{iql-i\alpha}re^{iql+i\alpha}r' + \cdots e^{iql-i\alpha}t' + \cdots$$
$$= \frac{trr'^2e^{3iql+i\alpha}}{1 - t'e^{iql-i\alpha}} \tag{1.87}$$

Four reflection terms

$$\psi_4^{tt}(l^-) = te^{iql+i\alpha}r'e^{iql-i\alpha}re^{iql+i\alpha}r'e^{iql-i\alpha}re^{iql+i\alpha} + \cdots t'e^{iql+i\alpha} + \cdots$$
$$= \frac{tr^2r'^2e^{5iql+i\alpha}}{1 - t'e^{iql+i\alpha}} \tag{1.88}$$

Five reflection terms

$$\psi_5^{tt}(l^-) = te^{iql+i\alpha}r'e^{iql-i\alpha}re^{iql+i\alpha}r'e^{iql-i\alpha}re^{iql+i\alpha}r' + \cdots e^{iql-i\alpha}t' + \cdots$$
$$= \frac{tr'^3r^2e^{5iql+i\alpha}}{1-t'e^{iql-i\alpha}} \tag{1.89}$$

So the contribution to the wavefunction inside the ring at the point $x = l^-$ coming from these diagrams will again form two geometric series to give,

$$\psi^{tt}(l^-) = [\psi_0^{tt}(l^-) + \psi_2^{tt}(l^-) + \psi_4^{tt}(l^-) + \cdots] + [\psi_1^{tt}(l^-) + \psi_3^{tt}(l^-) + \psi_5^{tt}(l^-) + \cdots]$$
$$= \frac{te^{iql+i\alpha}}{1-t'e^{iql+i\alpha}}(1 + r'r e^{2iql} + r^2r'^2e^{4iql} + \cdots)$$
$$+\frac{tr'e^{iql+i\alpha}}{1-t'e^{iql-i\alpha}}(1 + rr'e^{2iql} + r^2r'^2e^{4iql} + \cdots)$$
$$= \frac{te^{iql+i\alpha}}{1-t'e^{iql+i\alpha}}\frac{1}{1-rr'e^{2iql}} + \frac{tr'e^{iql+i\alpha}}{1-t'e^{iql-i\alpha}}\frac{1}{1-rr'e^{2iql}} \tag{1.90}$$

Thus, the resultant wavefunction at $x = l^-$ is (from Eqs. 1.83 and 1.90)

$$\psi(l^-) = \psi^{tb}(l^-) + \psi^{tt}(l^-)$$
$$\frac{t'}{1-t'e^{iql-i\alpha}}\frac{1}{1-rr'e^{2iql}} + \frac{t're^{2iql}}{1-t'e^{iql+i\alpha}}\frac{1}{1-rr'e^{2iql}}$$
$$+\frac{te^{iql+i\alpha}}{1-t'e^{iql+i\alpha}}\frac{1}{1-rr'e^{2iql}} + \frac{tr'e^{iql+i\alpha}}{1-t'e^{iql-i\alpha}}\frac{1}{1-rr'e^{2iql}}$$
$$= \frac{te^{iql+i\alpha}}{1-rr'e^{2iql}}[\frac{1}{1-t'e^{iql+i\alpha}} + \frac{r'}{1-t'e^{iql-i\alpha}}]$$
$$+\frac{t'}{1-rr'e^{2iql}}[\frac{1}{1-t'e^{iql-i\alpha}} + \frac{re^{2iql}}{1-t'e^{iql+i\alpha}}]$$

Note that this is again of the form $ae^{iql+i\alpha} + b^{-iql}$ where

$$a = \frac{t}{1-rr'e^{2iql}}[\frac{1}{1-t'e^{iql+i\alpha}} + \frac{r'}{1-t'e^{iql-i\alpha}}] \tag{1.91}$$

$$b = \frac{t'e^{iql}}{1-rr'e^{2iql}}[\frac{1}{1-t'e^{iql-i\alpha}} + \frac{re^{2iql}}{1-t'e^{iql+i\alpha}}] \tag{1.92}$$

Note that once we substitute $r' = re^{2iql}$ (explained below) then the expressions for a and b become the same in the absence of magnetic field. Also note that it will give the same current as that from Eqs. 1.74 and 1.75 at $x = 0^+$.

Note that in Fig. 1.1 the coordinate in the right lead is taken as x' with the $x' = 0$ point depicted in the figure. $x = 0$ point is also depicted. One can write the wavefunction on the right as

$$\psi(x \geq l) = te^{ik(x-l)} = te^{-ikl}e^{ikx} = t'e^{ikx} \tag{1.93}$$

In this picture, in the absence of potential $t' = 1$, as the wavefunction at $x = l$ should be e^{ikl}. So the entity $t = e^{ikl}$ in the absence of scatterer. So in the absence of the potential, the phase of t i.e., θ_t will be

$$\theta_t = kl$$

Therefore, its derivative with respect to incident energy E will be

$$\frac{d\theta_t}{dE} = \frac{d\theta_t}{dk}\frac{dk}{dE} = \frac{d(kl)}{dk}\frac{1}{\frac{dE}{dk}} = l\frac{1}{\frac{d}{dk}\frac{\hbar^2k^2}{2m}} = \frac{2ml}{\hbar^2}\frac{1}{2k} = \frac{ml}{\hbar^2 k} = \frac{2\pi l}{h}\frac{1}{\frac{\hbar k}{m}} = 2\pi l\frac{1}{hv}$$

The factor 2 comes because when we write the dispersion relation as $E = \frac{\hbar^2k^2}{2m}$ then both $\pm k$ states are taken into account. Therefore,

$$\frac{d\theta_t}{dE} = \pi\rho_0 \tag{1.94}$$

where $\rho_0 = \frac{1}{hv}$ per unit length is 1D DOS in the absence of scatterer. In the presence of scatterer, one can show (see Chap. 3)

$$\frac{d\theta_t}{dE} \approx \pi\rho \tag{1.95}$$

As is evident from Eq. 1.93 that

$$\theta_t = \theta_{t'} + kl \tag{1.96}$$

This naturally implies

$$\frac{d\theta_{t'}}{dE} \approx \pi(\rho - \rho_0) \tag{1.97}$$

We will work with $te^{ik(x-l)}$ or $te^{ikx'}$ means x' is a new coordinate system that is 0 at $x = l$. It is just a change of variable but it makes keeping track of phase changes with respect to the initial phase easier. In fact if the system has many such constituent parts joined by junctions, then it is suitable to assign a new origin for each region. However, the reflection and transmission at the different junctions in the

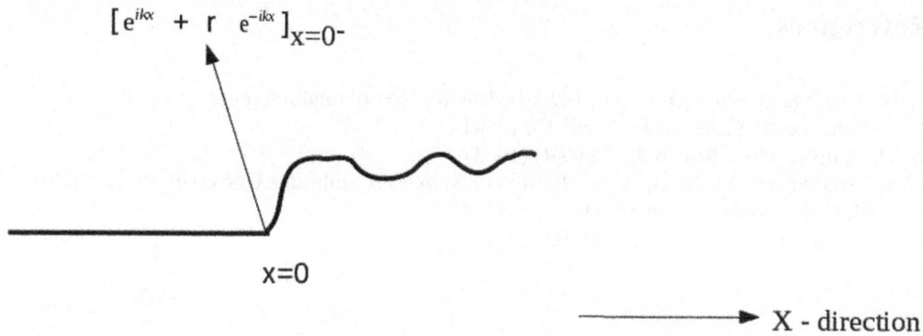

Fig. 1.15 The bold curve represents a 1D potential profile. x-direction is shown and choice of origin is also shown. This choice of origin implies the wavefunction that is matched at the origin is pointed with an arrow

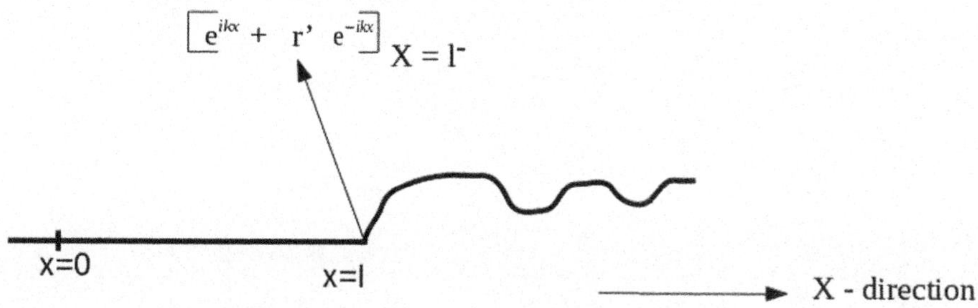

Fig. 1.16 The same scattering problem as that in Fig. 1.15 but the origin is shifted which changes the wavefunction that determines the boundary conditions

Feynman path approach has to be appropriately corrected for the necessary phase factor required. For example in Fig. 1.4 t is transmission amplitude of a three prong potential and r is the reflection amplitude. A simple calculation shows $t = 2/3$ and $r = -1/3$. This is the special case for the system solved in detail in Chap. 5. However, when the reflection occurs at $x = l$ then the reflection amplitude should be written as r' and it is related to r as follows:

$$\left\langle re^{-ik0^-} \middle| e^{ik0^-} \right\rangle = \left\langle r'e^{-ikl^-} \middle| e^{ikl^-} \right\rangle \tag{1.98}$$

This explains the substitution used in Eq. 1.76. r' is the reflection amplitude at $x = l$ while r is the reflection amplitude at $x = 0$. By symmetry, the amplitude of r and r' will be the same and equal to $-\frac{1}{3}$ but their phases will differ. This can be further understood by comparing Figs. 1.15 and 1.16. For the same reason, Eqs. 1.91 and 1.92 is explained since only anticlockwise moving $-k$ states are reflected at $x = 0$.

References

1. H. Bluhm et al., Phys. Rev. Lett. **102**(13), 136802 (2009). (and references therein)
2. G. Bonneau et al., Am. J. Phys. **69**, 322 (2001)
3. M. Buttiker, Phys. Rev. B **32**, 1846(R) (1985)
4. S. Datta, *Electronic Transport in Mesoscopic Systems* (Cambridge University Press, 1995)
5. S. Mukherjee et al., Physica E **118**, 113933 (2020)
6. S. Mukherjee et al., Physica E **44**, 62 (2011)

Chapter 2
Closed Systems

While, in the previous chapter, we introduced the reader to open systems with a simple example, that of a ring connected to an electron reservoir, in this chapter, we will present the closed version of the same system. While mesoscopic open systems are understood as a scattering problem, closed Hamiltonian systems are understood in terms of eigen states and eigen energies. Eigen states are connected to scattering states via Friedel sum rule that was preliminarily introduced in Chap. 1, Eq. 1.94. The simple principle behind arriving at the latter equation is that, by estimating scattering phase changes, we can figure out the eigen states of the system. Every π phase change correspond to an eigen state of the potential. The phase changes introduce new nodes in the eigen state and as the nodes increase, the potential can accommodate more and more eigen states. These facts will be explained in detail as we proceed. Thus, mesoscopic systems give us a route to make a one is to one connection between eigen states and states in an open system that exhibit dissipation and loss. Once dissipation and decoherence dominates the system becomes classical and new laws and principles become effective. Thus, we learn to treat classical systems and quantum systems as disjoint sets. The route to connect them can come from mesoscopic systems. We will continue with the simple one-dimensional (1D) ring for basic understanding and try to generalize in latter chapters.

A simple modification of the closed ring system will reveal some interesting aspects of the states of a closed systems that has only recently come to light. This incorporates a twist in the process by which eigen states evolve to become scattering states. In case of small quantum systems, FSR type of formulae will break down for two reasons. (1) Quantum corrections: they will be significant as these small quantum systems cannot be treated semi-classically. However, people have thought about it before us to some extent. (2) Number of states is one more than the number of nodes is not always true: this not only changes our view of the first point above but also leads to many dramatic consequences that require us to re-think the whole of physics in some sense. In Chap. 6, we will see that one can realize that signals can be sent in negative time (classically, this means faster than the speed of light) within the single particle coherence length.

P. Singha Deo, *Mesoscopic Route to Time Travel*,
https://doi.org/10.1007/978-981-16-4465-8_2

2.1 Parity Effect in Closed Ring

Consider an infinite square well potential of length L shown in Fig. 2.1. The quantum mechanical wavefunctions are drawn by dashed lines and they exhibit a parity effect wherein states with an even number of nodes and that with an odd number of nodes alternate. This gives an explicit example of the statement that the number labeling a state or the index of a state is one more than the number of nodes. Now we can also draw an infinite square well potential like a 1D line with perfectly reflecting dead ends. So the 1D line can also be curled up to form a ring as shown in Fig. 2.2. An Aharanov-Bohm flux can be placed in the center of the ring such that the electrons in the ring do not feel any magnetic field. The alternating even and odd state of the 1D line or the parity effect, manifests in the states of the ring in a particular way as we will discuss now.

Consider a 1D ring of length L. The quantum mechanical wavefunction in the ring is that of a free electron $\psi = e^{ikx}$. ψ is the solution to the 1D Schrodinger (Sc.) equation $-\frac{\hbar^2}{2m}\frac{d^2\psi}{dx^2} = E\psi$. The eigen energies can be found from the single valuedness of the wavefunction that yield the condition

$$\begin{aligned} e^{ik(x+L)} &= e^{ikx} \\ \text{or,}\ e^{ikL} &= 1 = e^{\pm i2n\pi} \qquad n = 0, 1, 2, \ldots \\ \text{or,}\ k &= \pm\frac{2n\pi}{L} \end{aligned} \tag{2.1}$$

$$E_n = \frac{\hbar^2 k_n^2}{2m} = \frac{2\hbar^2 n^2 \pi^2}{mL^2} \tag{2.2}$$

Now if an Aharanov-Bohm flux is introduced piercing the ring (note that there is no magnetic field $\vec{B} = \vec{\nabla} \times \vec{A}$ on the electrons in the ring as we keep the flux confined to a small area inside the ring), then the above condition changes to

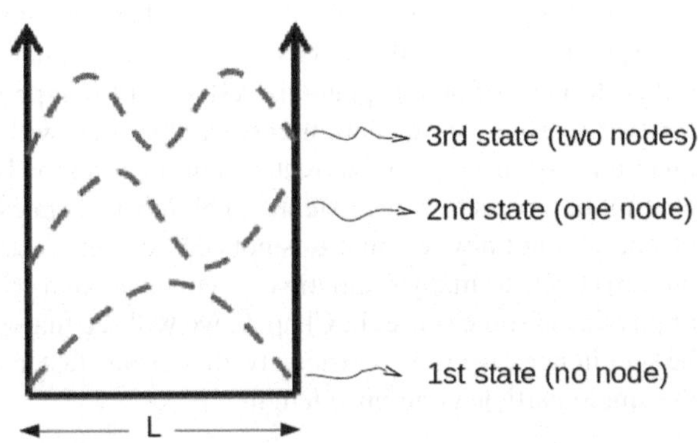

Fig. 2.1 An infinite potential well with states exhibiting parity effect

Fig. 2.2 A 1D ring of length L

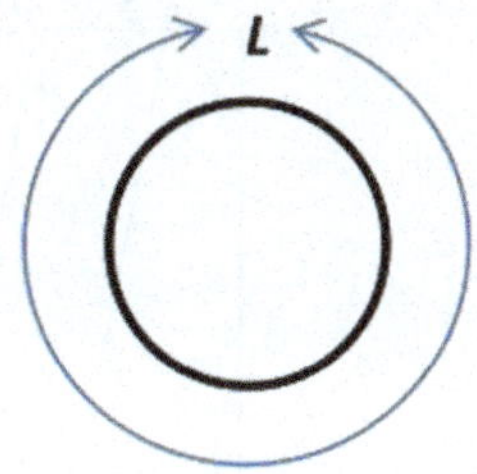

$$e^{i(kL-\alpha)} = 1 = e^{\pm i2n\pi} \qquad \text{where } \alpha = \frac{2\pi\Phi}{\Phi_0}$$

$$\text{or, } k = \frac{\pm 2n\pi + \alpha}{L} \tag{2.3}$$

where α can be positive as well as negative depending on the direction of the field.

$$\begin{aligned} E_n &= \frac{\hbar^2 k_n^2}{2m} = \frac{\hbar^2}{2m}\frac{(\pm 2n\pi + \alpha)^2}{L^2} \\ &= \frac{\hbar^2}{2mL^2}(\pm 2n\pi + \alpha)^2 \\ &= \frac{h^2}{2mL^2}\left(n + \frac{\Phi}{\Phi_0}\right)^2 \end{aligned} \tag{2.4}$$

where $n = 0, \pm 1, \pm 2, \ldots$.

The single particle levels E_n are drawn in Fig. 2.3. As is evident from Eq. 2.4 the single particle eigen energy E_n is parabolic in $\frac{\Phi}{\Phi_0}$. E_0 corresponding to $n = 0$ has minimum at $\frac{\Phi}{\Phi_0} = 0$ (see Eq. 2.4). E_1 corresponding to $n = +1$ is identical to E_0 but minimum displaced to $\frac{\Phi}{\Phi_0} = -1$. E_{-1} corresponding to $n = -1$ is identical to E_0 but minimum displaced to $\frac{\Phi}{\Phi_0} = +1$ and so on. The region between the two dotted lines, i.e., $|\frac{\Phi}{\Phi_0}| \le \frac{1}{2}$ defines the Brillouin zone as, by repeating it horizontally, we get rest of the spectrum. Now if we fix the flux $\frac{\Phi}{\Phi_0} = 1.34$, then the eigen energies of the ring are at the energies corresponding to the points P, Q, R, S, T, etc. The state at P is a diamagnetic state as its energy increases with flux and that at Q is a paramagnetic state as its energy decreases with flux. The slope of the single particle energy levels imply that they carry a current. The current carried by a particular level n is $I_n = -c\frac{\partial E_n}{\partial \Phi}$. This is a purely equilibrium current called persistent current and it occurs because the flux lifts the degeneracy between the clockwise moving and anticlockwise moving states. One can see that, at specific values of $\Phi = \frac{n}{2}\Phi_0$, these two time-reversed states become degenerate and their currents cancel each other. Thus, the paramagnetic and diamagnetic states alter just as in Fig. 2.1, the even and odd states alter. This is just the parity effect of the 1D line manifesting in a different way. States with an even number of nodes in the 1D line become diamagnetic in a 1D ring, while states with an odd number of nodes in the 1D line become paramagnetic

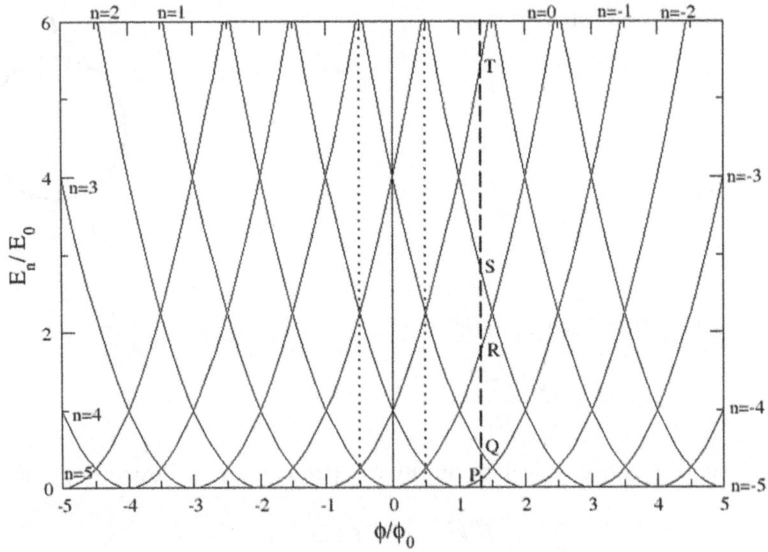

Fig. 2.3 The dotted lines encloses the first Brillouin zone. The dashed line is corresponding to $\frac{\Phi}{\Phi_0} = 1.34$

in the 1D ring. Counting of nodes can lead to very general arguments to state that even in the presence of many body effects and disorder, alternate many body states will carry opposite currents. These arguments are due to Leggett and outlined below.

Variational argument for parity effect in many electron states: The generalized electronic Hamiltonian for N electrons in a 1D ring, described by angular coordinate θ is,

$$H(\theta_1, \ldots \theta_N) = \sum_{i=1}^{N} V(\theta_i) + \sum_{i \neq j}^{N} U(\theta_i - \theta_j) + \sum_{i=1}^{N} \frac{\left[p_i - \frac{eA(\theta_i)}{c}\right]^2}{2m} \tag{2.5}$$

where, $V(\theta_i)$ is a one-body potential and $U(\theta_i - \theta_j)$ is a two-body potential. So, energy E is

$$\begin{aligned} E &= \langle \Psi(\theta_1, \ldots \theta_N) | H | \Psi(\theta_1, \ldots \theta_N) \rangle \\ &= \int \sum_i V(\theta_i) \Psi^*(\theta_1, \ldots \theta_N) \Psi(\theta_1, \ldots \theta_N) \, d\theta_1 \cdots d\theta_N \\ &+ \int \sum_{i \neq j} U(\theta_i - \theta_j) \Psi^*(\theta_1, \ldots \theta_N) \Psi(\theta_1, \ldots \theta_N) \, d\theta_1 \cdots d\theta_N + KE \qquad (2.6) \\ &= PE + KE \qquad (2.7) \end{aligned}$$

Here, KE refers to kinetic energy and PE consisting of the first two terms refer to potential energy. Now we want to construct a variational $\Psi(\theta_1, \ldots \theta_N)$. Let the variational wavefunction be

$$\Psi_{var}(\varphi, \theta_1, \ldots \theta_N)$$

We must ensure that Ψ_{var} respects the symmetries of the original Hamiltonian. So, Ψ_{var} should satisfy the following conditions.

1. If any one of $\theta_1, \theta_2, \ldots \theta_N$ change by 2π, Ψ_{var} should pick up a phase χ ($\chi = \frac{2\pi\Phi}{\Phi_0}$). Here, Φ is the total flux enclosed by the ring and $\Phi_0 = \frac{hc}{e}$. Note that, by this choice of Ψ_{var}, we are working with the gauge transformed Hamiltonian, where KE operator is $\frac{p_i^2}{2m}$ and not $\frac{(p_i - eA_i)^2}{2m}$. So,

$$\Psi_{var}(\varphi, \theta_1, \ldots \theta_N) = e^{i\chi(\theta_1, \ldots \theta_N)} \Psi(\varphi, \theta_1, \ldots \theta_N)$$

where,

$$-\chi(\theta_1, \ldots, \theta_i = 0, \ldots, \theta_N) + \chi(\theta_1, \ldots, \theta_i = 2\pi, \ldots, \theta_N) = \frac{2\pi\Phi}{\Phi_0} \tag{2.8}$$

2. Ψ_{var} should be anti-symmetric i.e.,

$$\Psi_{var}(\varphi, \theta_1, \theta_2, \ldots \theta_N) = -\Psi_{var}(\varphi, \theta_2, \theta_1, \ldots \theta_N) \quad \text{or for any other interchange of } \theta_i \text{ and } \theta_j.$$

Therefore,

$$e^{i\chi(\theta_1, \theta_2, \ldots \theta_N)} \Psi(0, \theta_1, \theta_2, \ldots \theta_N)$$

$$= -e^{i\chi(\theta_2, \theta_1, \ldots \theta_N)} \Psi(0, \theta_2, \theta_1, \ldots \theta_N)$$
$$= -e^{i\chi(\theta_2, \theta_1 \cdots \theta_N)} [-\Psi(0, \theta_1, \theta_2 \cdots \theta_N)] \tag{2.9}$$

as $\Psi(0, \theta_2, \theta_1 \cdots \theta_N)$ is a Fermionic wavefunction at $\varphi = 0$. So

$$\chi(\theta_1, \theta_2, \ldots \theta_N) = \chi(\theta_2, \theta_1 \cdots \theta_N) \text{ or for any other interchange of } \theta_i \text{ and } \theta_j \tag{2.10}$$

i.e., χ is a symmetric function of its indices.

As the initial phase is arbitrary, we choose $\Psi(0, \theta_1, \theta_2 \cdots \theta_N)$ to be real. For non-zero flux as well as for non-zero φ, Ψ_{var} is complex.

However, $\Psi_{var}(\Phi = 0)$ and $\Psi_{var}(\Phi \neq 0)$ have the same potential energy, as the potential energy depends on $|\Psi_{var}|^2$. Kinetic energy with these gauge transformed wavefunctions is

$$KE = \text{Re}\langle\Psi_{var}|\sum_i \frac{p_i^2}{2m}|\Psi_{var}\rangle \tag{2.11}$$

Thus,

$$KE = \text{Re}\int \Psi_{var}^* \sum_i \left(-\frac{\hbar^2}{2m}\right)\frac{\partial^2}{\partial\theta_i^2}\Psi_{var} d\theta_1 \cdots d\theta_N \tag{2.12}$$

In short, we write $\Psi_{var}(\Phi \neq 0) = \Psi_{var}(\Phi = 0)e^{i\chi} = \Psi_{or}e^{i\chi}$.

$$\begin{aligned}
KE &= -\frac{\hbar^2}{2m}\sum_i \text{Re}\int \Psi_{var}^* \frac{\partial}{\partial\theta_i}\frac{\partial}{\partial\theta_i}\Psi_{or}e^{i\chi} d\theta_1 \cdots d\theta_N \\
&= -\frac{\hbar^2}{2m}\sum_i \text{Re}\int \Psi_{var}^* \frac{\partial}{\partial\theta_i}\left[\Psi_{or} i\frac{\partial\chi}{\partial\theta_i}e^{i\chi} + e^{i\chi}\frac{\partial\Psi_{or}}{\partial\theta_i}\right] d\theta_1 \cdots d\theta_N \\
&= -\frac{\hbar^2}{2m}\sum_i \text{Re}\int \Psi_{var}^* \left[\Psi_{or} i\frac{\partial\chi}{\partial\theta_i} i\frac{\partial\chi}{\partial\theta_i}e^{i\chi} + \Psi_{or} i e^{i\chi}\frac{\partial^2\chi}{\partial\theta_i^2} + \frac{\partial\Psi_{or}}{\partial\theta_i} i\frac{\partial\chi}{\partial\theta_i}e^{i\chi}\right. \\
&\left. + e^{i\chi}\frac{\partial^2\Psi_{or}}{\partial\theta_i^2} + e^{i\chi} i\frac{\partial\chi}{\partial\theta_i}\frac{\partial\Psi_{or}}{\partial\theta_i}\right] d\theta_1 \cdots d\theta_N \\
&= -\frac{\hbar^2}{2m}\sum_i \text{Re}\int \left[-\Psi_{or}^*\Psi_{or}\left(\frac{\partial\chi}{\partial\theta_i}\right)^2 + \Psi_{or}^*\Psi_{or} i\frac{\partial^2\chi}{\partial\theta_i^2} + \Psi_{or}^* i\frac{\partial\chi}{\partial\theta_i}\frac{\partial\Psi_{or}}{\partial\theta_i}\right. \\
&\left. + \Psi_{or}^*\frac{\partial^2\Psi_{or}}{\partial\theta_i^2} + i\frac{\partial\chi}{\partial\theta_i}\Psi_{or}^*\frac{\partial\Psi_{or}}{\partial\theta_i}\right] d\theta_1 \cdots d\theta_N \\
&= -\frac{\hbar^2}{2m}\sum_i \int \left[-|\Psi_{or}|^2\left(\frac{\partial\chi}{\partial\theta_i}\right)^2 + \Psi_{or}^*\frac{\partial^2}{\partial\theta_i^2}\Psi_{or}\right] d\theta_1 \cdots d\theta_N
\end{aligned} \tag{2.13}$$

Therefore, change in kinetic energy due to Φ is

$$\begin{aligned}
&\frac{\hbar^2}{2m}\sum_i \int \left[|\Psi_{or}|^2\left(\frac{\partial\chi(\theta_1, \theta_2, \ldots \theta_N)}{\partial\theta_i}\right)^2\right] d\theta_1 \cdots d\theta_N \\
&= \frac{\hbar^2}{2m}N\int \left[|\Psi_{or}|^2\left(\frac{\partial\chi}{\partial\theta_1}\right)^2\right] d\theta_1 \cdots d\theta_N \geq 0
\end{aligned} \tag{2.14}$$

as the integrand is whole square. To explain the last step, we suggest as an example

$$\int |\Psi_{or}|^2\left(\frac{\partial\chi(\theta_1, \theta_2, \ldots \theta_N)}{\partial\theta_1}\right)^2 d\theta_1 \cdots d\theta_N + \int |\Psi_{or}|^2\left(\frac{\partial\chi(\theta_1, \theta_2, \ldots \theta_N)}{\partial\theta_2}\right)^2 d\theta_1 \cdots d\theta_N = J(\text{say})$$

In the second integration, we put $\theta_1 \longrightarrow \theta_2$ and $\theta_2 \longrightarrow \theta_1$, and we know that

$$\chi(\theta_1, \theta_2, \ldots \theta_N) = \chi(\theta_2, \theta_1 \cdots \theta_N)$$

Therefore,

$$J = \int |\Psi_{or}|^2 \left(\frac{\partial\chi(\theta_1, \theta_2, \ldots \theta_N)}{\partial\theta_1}\right)^2 d\theta_1 \cdots d\theta_N + \int |\Psi_{or}|^2 \left(\frac{\partial\chi(\theta_1, \theta_2, \ldots \theta_N)}{\partial\theta_1}\right)^2 d\theta_1 \cdots d\theta_N$$
$$= 2\int |\Psi_{or}|^2 \left(\frac{\partial\chi(\theta_1, \theta_2, \ldots \theta_N)}{\partial\theta_1}\right)^2 d\theta_1 \cdots d\theta_N$$

This explains why the ground state, second excited state or any even state energy always increases with flux (for $\Phi < \Phi_0/2$) that is it is a diamagnetic state. The first excited state however does not or to be precise does not necessarily do so. This follows from the following topological arguments. Suppose we have a ring with an even number of electrons N. Now let us fix the coordinates of $N-1$ electrons which is an odd number of electrons and take the Nth electron round the ring once. It has to pick up a phase of 2π when it comes back to its initial position (note that this is a mathematical operation that does not care whether one can fix the position of other electrons and whether the test electron can be physically taken round back to its exact initial position). Every time it crosses an electron, it picks up a phase of π due to anti-symmetric property of the many body wavefunction. So when it goes round the ring once, it will pick up a π phase an odd number of times compared to an even number of times if we started with an odd number of electrons in the ring. This extra phase difference of π for the many body energy of even number of electrons with respect to that for odd number of electrons do not necessarily change continuously with any parameter like χ. So the first excited state, third excited state or any odd state needs a non-symmetry-dictated node (NSDN) to match its wavefunction in going round the ring once and this makes them paramagnetic. To be precise, any arbitrary function on either sides of a NSDN is automatically matched by the NSDN and so the odd electron states are at least non-diamagnetic.

2.2 Break Down of Parity Effect

Now let us consider two infinite square well potentials shown in Fig. 2.4 placed perpendicular to each other (well in topology, the exact angle of meet between the two lines or intervals does not matter because topological spaces do not necessarily have a notion of distance and angle). We will indeed see that the quantum mechanical equations do not depend on the angle. However, in latter chapters, when we show physical realizations of this system curved out of real three-dimensional (3D) quantum wires then the two lines better be perpendicular for theoretical convenience and such systems has been fabricated in the laboratory confirming the results to be discussed below (Kobayashi et al. 2004). One is shown by the solid line AB and the other by the solid line CD. They are decoupled from each other in Fig. 2.4a. The

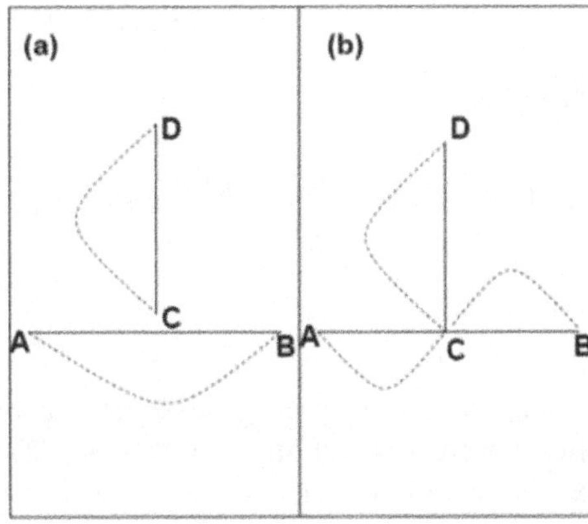

Fig. 2.4 Two infinite potential wells are place perpendicular to each other. In **a**, they are disjoint and in **b**, they are joined

ground state wavefunction in AB and CD are shown by the dotted lines. Both are even states. In Fig. 2.4b, we join AB and CD. It is to be noted that this kind of topological spaces are well studied in algebraic topology and differential geometry as graphs and algebraic varieties. Quantum phenomenon in such spaces is a motivation induced by mesoscopic physics and the ability to make such samples in the laboratory. CD now forces a node at the point where it touches AB. The ground state along AB now has a node and it is a NSDN that introduces an additional phase change of π. In contrast to that of Leggett this is a physical NSDN and will have physical consequences to be discussed gradually. We will first show that it leads to a violation of parity effect. The phase change for NSDN in the variational wavefunction of Leggett does not produce a breakdown of parity effect. The origin of phase changes in a 1D ring are phase change due to propagation, Aharonov-Bohm effect and anti-symmetry of many body wavefunction. Leggett used the NSDN in his variational wavefunction to count the phase change due to propagation and said that there will be an integral times π difference between N body wavefunction and $N+1$ body wavefunction. Let us first try to understand the role of this physical NSDN on parity effect and its consequences. For this, we can join the two ends A and B to form a ring as shown in Fig. 2.5. We can now put an Aharanov-Bohm flux through the center of the ring to get a persistent current or magnetization in the ring. The length CD can be varied continuously so that it does not make an exact node on AB, in which case the states in CD will leak into the ring formed by closing AB and the question is whether these states will be paramagnetic or diamagnetic? The phase change of π seen in Fig. 2.4 is not related to propagation, or Aharonov-Bohm effect or anti-symmetry of wavefunctions.

One can see Singha Deo (1996) where we solve for a situation wherein length v can be many times longer than length u. Which means the spacings between the levels of the stub will be much smaller than the levels in the ring. Now when the levels of the stub leak into the ring as they are joined then they will become magnetic and they will violate the parity effect. There can be many consecutive states that are all diamagnetic (or paramagnetic). Below, we present the basic calculations and facts. Let us solve for the currents in the system shown in Fig. 2.5. The wavefunctions in

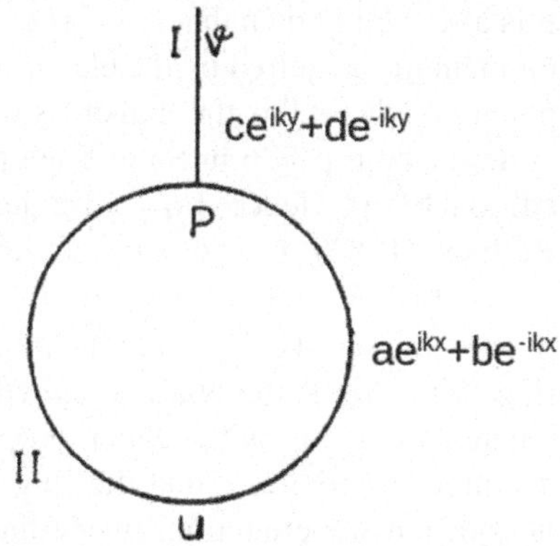

Fig. 2.5 A ring attached to a finite side branch referred to as stub

the different regions is also shown in Fig. 2.5. The origin is chosen at the junction point P and the boundary conditions yield the following equations:

$$a + be^{-i\alpha} = ae^{iku+i\alpha} + be^{-iku} = c + d$$
$$ik(a - be^{-i\alpha}) - ik(ae^{iku+i\alpha} - be^{-iku}) + ik(c - d) = 0$$
$$ce^{ikv} + de^{-ikv} = 0$$

Here, $\alpha = 2\pi\Phi/\Phi_0$ is the AB phase, Φ being the flux through the ring and Φ_0 is the flux quantum. From these equations, one can solve for the allowed modes in the system and is given by the following condition:

$$cos(\alpha) = \frac{sin(ku)cot(kv)}{2} + cos(ku) \tag{2.15}$$

Solving for k, we get the allowed wavevectors. One can check that the above condition is the simplified form of the following condition.

$$cos(\alpha) = re(1/T) \tag{2.16}$$

where T is the transmission amplitude across a wire of length u with an attached side arm of length v. This is the structure we get when the ring in Fig. 2.5 is cut open and formed into a 1D line, i.e., we get the system shown in Fig. 2.4b. The above equation immediately suggests the following. For the clean ring, the bound state condition is $e^{i(ku-\alpha)} = 1$ (see Eq. 2.3), whereas Eq. 2.16 is just the condition (It is worth mentioning that $cos(\alpha) = cos(2n\pi - \alpha)$)

$$e^{icos^{-1}(Re(\frac{1}{T})-\alpha)} = 1 \tag{2.17}$$

Buttiker et al. (1983) have shown that an electron in a ring with a random potential is effectively moving in a periodic potential whose unit cell is the ring when cut open.

It is also well known that $cos^{-1}(re\frac{1}{T})$ is the Block phase (Ku where K is the Block momentum) acquired by the electron in traversing an unit cell of an infinite periodic potential where T is the transmission amplitude across the unit cell of the periodic system (see Eq. 8.76 in Solid State physics by Ashcroft and Mermin, International edition 1976). Hence, Eq. 2.17 is just $e^{i(Ku-\alpha)}$=1 and is in perfect agreement with Buttiker (1983). It also suggests that inside the ring the electron moves with the momentum K and not with the free particle momentum $\pm k$, although the energy of the state will be $\frac{\hbar^2 k^2}{2m}$, and the k values can be obtained from Eq. 2.15. In a clean ring, $(ku - \alpha)$ is the phase acquired by the electron in going round the ring once, whereas in a ring with scatterers (potential or geometric), $(Ku - \alpha)$ is the phase acquired in moving round the ring once and Eq. 2.17 is due to the single valuedness of the wavefunction. Inside the ring, the electron is moving anticlockwise or clockwise with momentum $\pm K$(where $Ku = cos^{-1}(re\frac{1}{T})$). One of them is a diamagnetic (anticlockwise moving) state and the other is a paramagnetic (clockwise moving) state. Initially, as the magnetic field is increased, the two states ($\pm K$) move away from each other; however, the diamagnetic state and the paramagnetic state are not degenerate for any value of Φ as the scatterer breaks the rotational symmetry of the ring and lifts the degeneracy between the clockwise and anti-clockwise states at $\Phi = n\Phi_0/2$.

So the bound states can be determined by graphically solving Eq. 2.16 or $re(1/T) = cos(\alpha)$. So in Fig. 2.6, we show a simple plot where the solid curve is a plot of $y = re(1/T)$ with ku for $v/u = 0.2$. Wherever this curve intersects the straight line $y = cos(\alpha)$, the corresponding k value gives a bound state for the system for a particular value of α. Let us start with $\alpha = 0$ and then $y = cos(0)$ curve is shown in Fig. 2.6 by dotted straight lines. Two consecutive points where the curve $y = re(1/T)$ intersects the straight line $y = cos(0)$ are denoted by A and B in the

Fig. 2.6 Dispersion curve for the system in Fig. 2.5 for $v/u = 0.2$

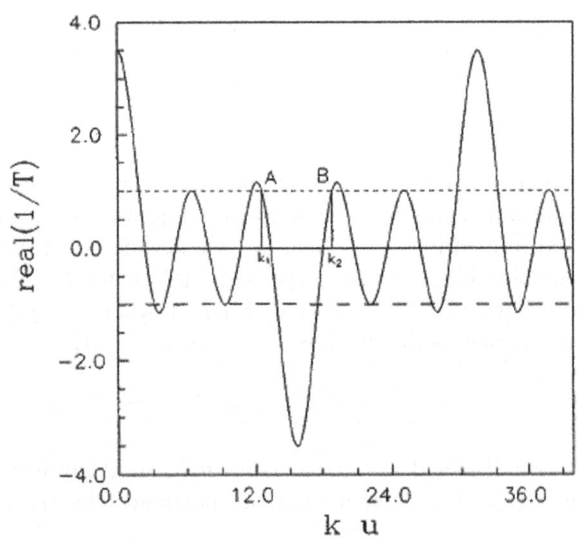

figure. The corresponding k values are denoted as k_1 and k_2 in the figure. For $\alpha = 0$, we can get two of the allowed modes of the system by putting the values k_1 and k_2 in Eqs. 2.2 and 2.17. If α is increased gradually then the dotted straight line $y = cos(\alpha)$ shifts gradually downwards towards the dashed straight line. As the curve $y = cos(\alpha)$ gradually go downwards, the allowed wavevectors k_1 and k_2 slowly drift rightwards and leftwards, respectively, along the k-axis. As k_1 drifts rightwards with α, i.e., towards higher energy, k_1 is a diamagnetic state. Similarly k_2 is a paramagnetic state. That k_1, k_2, etc., gradually increase or decrease with α gives rise to a dispersion with α (E vs α) with close by alternate states going further away from each other with α up to $\alpha = \pi$. $y = cos(\pi)$ is also shown in Fig. 1 as a dashed line. If we increase α further then the straight curve $y = cos(\alpha)$ start moving upwards and comes back to its original position at $\alpha = 2\pi$. This ensures Φ_0 periodicity of the dispersion curves. Since $cos(\alpha)$ can vary from -1 to $+1$ (dotted lines to dashed lines), the dispersion curve for any two consecutive states can never cross (see Fig. 2.6). So the dispersion curve is exactly similar to that of a ring with a random potential. The cause of gaps in the dispersion curve in that case is the breakdown of rotational symmetry of the ring by the random potential and hence the removal of degeneracy of states that cross for a clean ring. In our case, the rotational symmetry is destroyed by the topological defect as compared to potential scattering considered by Buttiker (1983). Some gaps (the ones around $ku = 2n\pi$) are very large, but most of the gaps are very small. In fact, some special gaps may actually go to zero for reasons explained later. Hence, from Fig. 2.6, it is evident that alternate states carry persistent currents with opposite signs and have opposite magnetic properties up to infinite energy. This is exactly the same as in case of potential scattering.

The effects are different when we plot the same curves for different values of v/u = 0.21 (see Fig. 2.7) (in fact $v/u = 0.2 \pm \epsilon$ is sufficient). Consider the intersections

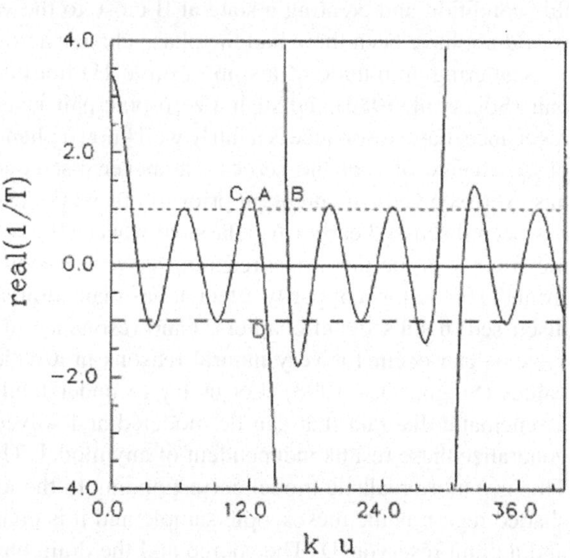

Fig. 2.7 Dispersion curve for the system in Fig. 2.5 for $v/u = 0.21$

between the curves $y = re(1/T)$ and $y = cos(0)$. The first few alternate states have opposite magnetic properties, but the fifth and the sixth states (two consecutive states marked A and B) are both diamagnetic disobeying the parity effect. One can slowly increase α to see that. Parity effect is again violated for the 11th and the 12th states both of which are paramagnetic. After a regular spacing of five levels we always find two consecutive levels that violate the parity effect. We shall soon see why it does not happen for specific values of $v/u = 0.2$ (say).

A special feature of the delta potential is that $| T |^2 = re(T)$. This feature is not seen for any other potential. However, this feature is also observed in case of a stub. This makes it possible to map a single stub to an effective delta potential $V(x) = kcot(kv)\delta(x)$. So the strength of the delta potential depend on the Fermi energy. That is why, at certain energies, $V(x)$ become zero and then the gaps vanish. For $k = 0$ $V(x) = 1/v$, which means it starts with a small positive value. Then it decreases and soon goes to zero. After this, the strength of the potential monotonously increase on the negative side and finally becomes $-\infty$ at $kv = \pi$. After this, $V(x)$ undergoes a discontinuous jump from $-\infty$ to $+\infty$. If the strength of the δ potential at $kv = \pi - \epsilon$ and $kv = \pi + \epsilon$ are discontinuous the scattering phase shift and hence the Block phase will also undergo discontinuous jump. $re(1/T)$ also make a discontinuous jump and hence the Block phase jumps by π (see Fig. 2.7) as the wavevector k or energy is increased (Block phase of the infinite periodic system has to be defined to a modulo of 2π i.e., $-1 < re(1/T) < 1$). The next allowed Block phase of the infinite periodic system of stubs after that at D is that at B and they differ by π. This is markedly different from the next allowed Block phase at any other gap, e.g., the Block phase at C is same as that at A. We have seen that if the Block phase of the periodic system of stubs equals the AB phase α (see Eq. 2.16. For the time being, we have taken $\alpha = 0$) the single valuedness of the wavefunction gets satisfied in the ring and we get a bound state. This additional discontinuous π phase results in satisfying this condition and creating a state at B close to the value $kv = \pi$ which otherwise would not have been there had the phase change across kv = π been continuous.

Scattering amplitude of a stub (a finite 1D line) is characterized by a zero-pole pair (Sols et al. 1989) and such a zero-pole pair is a characteristic feature of Fano resonance. Fano resonance is a fairly well known phenomenon, but the discontinuous phase change of π at the zeroes was never discussed and no manifestations of it has ever received any attention prior to our works (Singha Deo 1996). Although it has been discussed earlier for classical waves (Berry 1998), its role in determining mesoscopic properties are far reaching as we will show in subsequent chapters. Apart from the breakdown of parity effect, it has many dramatic manifestations that will be discussed in subsequent chapters. Fano resonance in low-dimensional mesoscopic systems can occur for very natural reasons in a wide range of realistic parameter values (Singha Deo 1998). Let us try to understand this further with the help of a schematic diagram that can be modeled and solved. In Chaps. 5 and 6, we will generalize these results independent of any model. The system is shown in Fig. 2.8. This is a more realistic model corresponding to the ideal situation in Fig. 2.4b. The shaded region is the mesoscopic sample and it is connected to a source reservoir S and a drain reservoir D. The source and the drain have chemical potentials μ_1 and

μ_2 as well as temperatures T_1 and T_2, respectively, as shown in the figure. However, these parameters along with the Hamiltonian of the sample are not enough to tell us anything about the properties of the system. We also need to consider the leads L and R that connect the sample to the reservoirs. The exchange of energy and particles between the system and the reservoirs take place through the leads and the nature of the leads along with the points where it connects to the sample are very important and are well accounted for in the Landauer-Buttiker formalism. If the leads are connected to different points then the properties of the system will change completely. If the confinement potential of the leads are different then the nature of the states in the leads will be different and again that will change the properties of the system. This is the reason why intrinsic length scales of the sample (shaded region) are not dominant and does not lead to material specific parameters like resistivity. This confinement potential in the leads act in the transverse direction, while in the longitudinal direction from S to D, the states are plane wave propagating states. In the figure, the transverse confinement is taken to be hard wall which makes the transverse states like the states in a 1D infinite well (shown in the leads in Fig. 2.8). These will be mathematically shown in Chap. 4, and here we just provide some motivation based on physical arguments. A few such transverse states are shown in the left lead L as well as in the right lead R. If the transverse confinement is taken to be harmonic, then these states will be different and properties of the system will change. The two leads are taken to be identical and in a real situation one lead may be thinner than the other. Again that will change the system properties drastically. One can connect the system to more than or less than two reservoirs as is the case for making a four probe measurement versus two probe measurement. In simple words, the leads are an integral part of the system and has to be accounted for explicitly in terms of the positions where they are attached, exact number of them that are attached, nature of the transverse states in them, etc. When μ_1 and μ_2 are different then we get a non-equilibrium situation and a transport current will flow through the system. When they are the same the system will have typical equilibrium thermodynamic properties. Same holds for T_1 and T_2. The thermodynamic and transport properties of such a grand canonical system can only be understood from the scattering matrix elements for electrons being incident from the source reservoir. These electrons can be transmitted to the drain reservoir or reflected back to the source reservoir. The transverse mode of incidence and the transverse mode to which it is scattered crucially determine the scattering amplitudes. Depending on the experimental situation we have to consider the particular modes involved and an averaging over the modes is not a freedom that we have. In the linear response regime the same scattering matrix that determine the thermodynamic properties also determine the transport properties but the relevant matrix elements vary. In a situation where the sample dimension is less than the inelastic mean free path, only elastic scattering has to be considered. Otherwise, one has to use a suitable model to account for inelastic scattering. But the crucial matter of importance is that only these scattering matrix elements include the effect of the nature of modes in the leads and the nature in which the leads connect to the system. Simply, the Hamiltonian of the system or its free energy does not include these effects that are the most important to understand the mesoscopic properties.

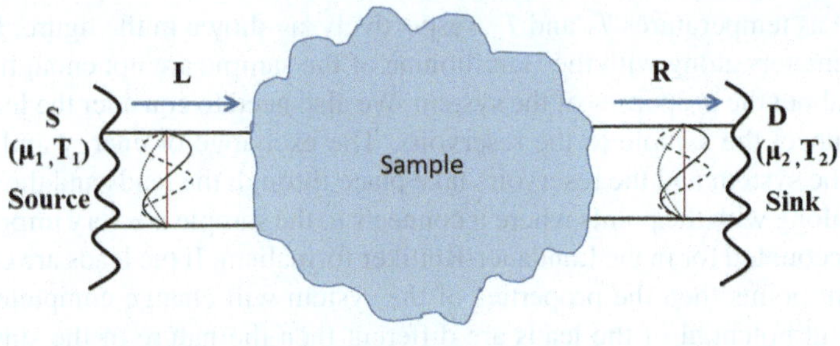

Fig. 2.8 Schematic diagram of a general mesoscopic system

References

1. M.V. Berry, J. Mod. Opt. **45**, 1845–1858 (1998)
2. M. Buttiker, Y. Imry, R. Landauer, Phys. Lett. **96A**, 365 (1983)
3. K. Kobayashi, H. Aikawa, S. Katsumoto, Y. Iye, Phys. Rev. B **68** 235304 (2003); K. Kobayashi, H. Aikawa, A. Sano, S. Katsumoto, Y. Iye, Phys. Rev. B **70** 035319 (2004)
4. P. Singha Deo, Phys. Rev. B **53**, 15447 (1996)
5. P. Singha Deo, Solid St. Commun. **107** 69 (1998)
6. F. Sols et al., J. Appl. Phys. **66**, 3892 (1989)

Chapter 3
Larmor Clock and Friedel Sum Rule

In this chapter, we present a systematic derivation of Friedel sum rule for mesoscopic systems which is quite different from that used in condensed matter physics in the sense that it consists of a hierarchy of formulas rather than a single formula. It eventually leads to a similar relation between DOS and scattering phase shifts, but with important correction terms to the formula presented in condensed matter text books (Ziman 1979). In Chaps. 5 and 6, we will show that the phase lapses discussed in Chap. 2 further complicates the hierarchy of formulas, in fact in a beneficial manner.

3.1 Larmor Precession Time

In this section, we present a systematic derivation of the hierarchy of formulas based on the classical concept of Larmor precision. It is an elaboration of the proof presented by Buttiker (2001). There he also presented several formulas that relate several physical observables like AC response, non-linear response, decoherence, etc., based on this hierarchy of formulas, which however, has not attracted much attention so far. We will derive only the hierarchy of formulas as that is rather basic. We will further elaborate these formulas in the next few chapters with respect to the phase lapses discussed in Chap. 2. Hopefully, that will motivate researchers in studying mesoscopic response in greater details.

We consider an arbitrary potential $V(y)$ for an electron (or any quantum particle) propagating in the y-direction. Asymptotically (i.e., $|y| \to \infty$) $V(y) = 0$. Therefore, this defines a one-dimensional (1D) scattering problem and explained in more details in the caption of Fig. 3.1. The incident electron stationary beam has a spin polarized in the x-direction. This stationary beam consists of an ensemble of particles (a quantum ensemble is different from a mesoscopic grand canonical ensemble). Some spins point in the positive x-direction and some in the negative x-direction resulting in a net spin in positive x-direction. Similarly, some spin point in positive z-direction

P. Singha Deo, *Mesoscopic Route to Time Travel*,
https://doi.org/10.1007/978-981-16-4465-8_3

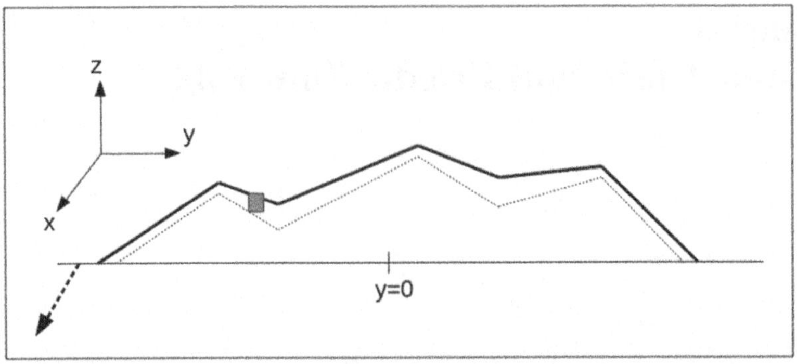

Fig. 3.1 Dotted line represents scattering potential $V(y)$. Thick solid line represents potential $V(y) + \delta V(y)$, where $\delta V(y)$ is a very small functional change in potential due to a magnetic field in the z-direction. A differential change in potential $dV(y')$ at $y = y'$ is shown by the shaded square block between the dotted and dashed lines. The area of this block $dV(y')dy'$ is therefore an integration measure. The dashed arrow represents incident electron beam spin polarized in x-direction. The axis directions and origin marked $y = 0$ are shown in the figure. As usual the region to the left of the potential is the left lead that can be labeled β and that to the right can be labeled α. So the incident electron or spin is coming from β and going towards α

and some in negative z-direction with no net spin in z-direction. Similarly, there is no net spin in y direction. In the absence of magnetic field $\vec{B}$, the scattering matrix elements $s_{\alpha\beta}(E, V(y))$ for spin up as well as spin down electrons is a function of energy and functional of potential. Here, α and β are the indices for outgoing and incoming asymptotic modes. Now suppose there is a small non-uniform magnetic field $B(y)$ applied in the z-direction in the region where the potential $V(y) \neq 0$. When magnetic field is applied, magnetic moments experience a torque because of which they precess about the magnetic field. This is called Larmor precession. If $\vec{\mu}$ is the magnetic moment of a particle in a magnetic field $\vec{B}$, then the torque experienced is

$$\vec{\tau} = \vec{\mu} \times \vec{B} = \gamma(\vec{J} \times \vec{B}) \tag{3.1}$$

$\vec{J}$ is the angular momentum vector of a spin $\frac{1}{2}$ electron, which will precess with angular frequency ω_L called Larmor frequency. This torque $\vec{\tau}$ is therefore in the y-direction (meaning force in z-direction) in the situation depicted in Fig. 3.1 and so the transmitted (or reflected) electrons will have a spin with a y-component as well. The magnetic field creates the potential $\delta V(y)$ on top of the potential $V(y)$. So for spin up (spins pointing in positive z-direction) electrons the potential energy is reduced by $\gamma\mu B(y) = \hbar\frac{\omega_L}{2}$ and for spin down electrons the potential energy is increased by $\hbar\frac{\omega_L}{2}$ where $\omega_L = \frac{2\gamma\mu B}{\hbar}$. This defines $\delta V(y) = \mp\gamma\mu B(y)$, where γ is the gyromagnetic ratio. This potential energy will try to orient the spins in z-direction but this will happen through precessional motion or Larmor precession. So the transmitted electrons will develop a spin with a z-component as well as y-component. We will concentrate here on the y-component as it helps us address the problem of density of states.

Thus, the scattering matrix element in the presence of magnetic field, (for small field) can be expanded as

$$s^{\pm}_{\alpha\beta}(E, V(y) \mp \delta V(y)) = s_{\alpha\beta}(E, V(y)) \mp \int dy' \frac{\delta s_{\alpha\beta}(E, V(y'))}{\delta V(y')} \delta V(y') + ... \tag{3.2}$$

$\frac{\delta s_{\alpha\beta}(E,V(y'))}{\delta V(y')}$ is a functional derivative. This is a general situation which in a special case for a differential change in potential at the point $y = y'$ can be written as

$$s^{\pm}_{\alpha\beta}(E, V(y') \mp dV(y')) = s_{\alpha\beta}(E, V(y')) \mp \left[\frac{ds_{\alpha\beta}(E, V(y'))}{dV(y')}\right] dV(y')dy' + ... \tag{3.3}$$

The asymptotic states is a natural choice for the basis states of a scattering problem which in this case is a spinor. The basis spinor which we use to calculate the precession angle of transmitted electrons is given by

$$|\psi_2\rangle = \begin{bmatrix} \psi_{2+} \\ \psi_{2-} \end{bmatrix} = \frac{1}{\sqrt{|s^+_{21}|^2 + |s^-_{21}|^2}} \begin{bmatrix} s^+_{21} \\ s^-_{21} \end{bmatrix}$$

Here, the incoming index is $\beta = 1$ and outgoing index is $\alpha = 2$. For example, $\beta = 1$ and $\alpha = 1$ can be used to do the same analysis for reflected electrons, while $\beta = 2$ sector will correspond to electrons incident from the right. Here, $|\psi_2\rangle$ is transmitted wavevector, ψ_{2+} is the spin up component and ψ_{2-} is the spin down component in transmitted channel 2. Here, s^+ and s^- are the scattering matrix for spin up and spin down electrons. For $1 \rightarrow 2$ scattering, the y-component of spin is

$$\langle s^y_{21}\rangle = \langle\psi_2|\frac{\hbar}{2}\sigma^y|\psi_2\rangle \tag{3.4}$$

where σ^y is y-component of Pauli matrices.

$$\langle s^y_{21}\rangle = \frac{\hbar}{2}\frac{1}{|s^+_{21}|^2 + |s^-_{21}|^2} \begin{bmatrix} s^{+*}_{21} & s^{-*}_{21} \end{bmatrix} \begin{bmatrix} 0 & -i \\ i & 0 \end{bmatrix} \begin{bmatrix} s^+_{21} \\ s^-_{21} \end{bmatrix}$$

$$= \frac{\hbar}{2}\frac{1}{|s^+_{21}|^2 + |s^-_{21}|^2} \left(-is^{+*}_{21} s^-_{21} + is^{-*}_{21} s^+_{21}\right)$$

$$= \frac{-i\hbar}{2}\frac{1}{|s^+_{21}|^2 + |s^-_{21}|^2} \left(s^{+*}_{21} s^-_{21} - s^{-*}_{21} s^+_{21}\right)$$

Using the expansion of $s_{21}^{\pm}$ from Eq. 3.2, we get

$$\langle s_{21}^{y}\rangle = \frac{-i\hbar}{2\left(|s_{21}^{+}|^2+|s_{21}^{-}|^2\right)}\left[\left(s_{21}^{*}-\int dy'\frac{\delta s_{21}^{*}}{\delta V(y')}\delta V(y')\right)\left(s_{21}+\int dy'\frac{\delta s_{21}}{\delta V(y')}\delta V(y')\right)\right]$$
$$+\frac{i\hbar}{2\left(|s_{21}^{+}|^2+|s_{21}^{-}|^2\right)}\left[\left(s_{21}^{*}+\int dy'\frac{\delta s_{21}^{*}}{\delta V(y')}\delta V(y')\right)\left(s_{21}-\int dy'\frac{\delta s_{21}}{\delta V(y')}\delta V(y')\right)\right]+\cdots$$

Multiplying and retaining the terms up to first order in $\delta V(y')$, we get

$$\begin{aligned}\langle s_{21}^{y}\rangle &= \frac{-i\hbar}{2T}\left[\int dy' s_{21}^{*}\frac{\delta s_{21}}{\delta V(y')}\delta V(y') - HC\right]\\ &= \frac{-i\hbar}{2T}\left[\int dy'\left(s_{21}^{*}\frac{\delta s_{21}}{\delta V(y')} - HC\right)\delta V(y')\right]\end{aligned}$$

where $T=2[|s_{21}^{+}|^2+|s_{21}^{-}|^2]$ and HC stands for Hermitian conjugate. Consider that the magnetic field is constant in a small interval $[y, y+dy]$ and takes there the value B. $\delta V(y)$ is the perturbation in potential due to the presence of magnetic field, hence, $\delta V(y)$ vanishes everywhere except in the interval $[y, y+dy]$, where it takes the value $dV=\frac{\hbar\omega_L}{2}$. The potential is reduced by $\frac{\hbar\omega_L}{2}$ for spin up electrons and increases by $\frac{\hbar\omega_L}{2}$ for spin down electrons. Therefore, in this interval, we can drop the integral and then the y-component of spin is given by

$$\langle s_{21}^{y}\rangle = \frac{-i\hbar}{2T}\left[dy'\left(s_{21}^{*}\frac{\delta s_{21}}{\delta V(y')} - HC\right)\frac{-\hbar\omega_L}{2}\right]$$

Therefore, at point y' it is given by,

$$\langle s_{21}^{y}\rangle = \frac{-i\hbar}{2T}\left[\left(s_{21}^{*}\frac{\delta s_{21}}{\delta V(y)} - HC\right)\frac{-\hbar\omega_L}{2}\right]$$

For unit spin, the y-component developed is given by

$$\langle s_{21}^{y}\rangle \equiv \frac{\langle s_{21}^{y}\rangle}{\frac{\hbar}{2}} = \frac{i\hbar}{2T}\left[\left(s_{21}^{*}\frac{\delta s_{21}}{\delta V(y)} - HC\right)\omega_L\right] \tag{3.5}$$

y-component of unit spin is equivalent to angle of precession of the electrons. Dividing this quantity by Larmor frequency ω_L, then give a time which is called Larmor precession time and is given by

$$\begin{aligned}\tau_{21}^{y}(y',E) &= \frac{i\hbar}{2T}\left[\left(s_{21}^{*}\frac{\delta s_{21}}{\delta V(y)} - HC\right)\right]\\ &= \frac{-h}{4\pi i T}\left[\left(s_{21}^{*}\frac{\delta s_{21}}{\delta V(y)} - HC\right)\right]\end{aligned} \tag{3.6}$$

Note that we are dividing a quantity calculated purely from quantum mechanics by the classical Larmor frequency. This is similar to Landau eigen energy divided by $\hbar$ giving cyclotron frequency $\frac{eB}{mc}$. The same analysis can be repeated in quasi-one dimension (which will be described in detail in the next chapter) to get the time spent by an electron going from β to α at point $\mathbf{r}$ (which obviously means we can always define a physical quantity by averaging over $\mathbf{r}$) within the scattering region as given by

$$\tau_{\alpha\beta}(\mathbf{r}, E) = \frac{-h}{4\pi i |s_{\alpha\beta}(E)|^2} Tr\left(s_{\alpha\beta}(E)^\dagger \frac{\delta s_{\alpha\beta}(E)}{\delta V(\mathbf{r})} - s_{\alpha\beta}(E)\frac{\delta s_{\alpha\beta}(E)^\dagger}{\delta V(\mathbf{r})}\right) \quad (3.7)$$

Tr is trace which represents summation over incident and outgoing transverse mode (channel) indices which are generally momentum indices, while α and β are spatial indices corresponding to the spatial positions where the leads attach to the sample (which will again become clearer in the next chapter). For multiple leads, β and α can take values 1, 2, 3, ... From Larmor precession time, a hierarchy of partial density of states (DOS) is introduced in Buttiker (2001). Although the derivation so far seems semi-classical, it yields exact agreement with quantum mechanical calculations apart from situations we will bring out subsequently and we will also provide a general understanding based on topological arguments.

3.2 Hierarchy of Mesoscopic Formulas

We now change the notation in order to discuss the hierarchy of relations between scattering phase shifts and DOS to write Larmor precession time as

$$\tau_{lpt}(E, \alpha, \beta, \mathbf{r}) = -\frac{\hbar}{4\pi i |s_{\alpha\beta}(E)|^2}\left[s^*_{\alpha\beta}\frac{\delta s_{\alpha\beta}}{e\delta U(\mathbf{r})} - \frac{\delta s^*_{\alpha\beta}}{e\delta U(\mathbf{r})}s_{\alpha\beta}\right] \quad (3.8)$$

Electrons in an ensemble are indistinguishable fermions that occupy one state each at zero temperature. Thus, we define local partial density of states (LPDOS) ρ_{lpd} in the same sense as it is defined on a 1D line as $\frac{|s_{\alpha\beta}(E)|^2}{\hbar}\tau_{lpt}$.

$$\rho_{lpd}(E, \alpha, \beta, \mathbf{r}) = -\frac{1}{4\pi i}\left[s^*_{\alpha\beta}\frac{\delta s_{\alpha\beta}}{e\delta U(\mathbf{r})} - \frac{\delta s^*_{\alpha\beta}}{e\delta U(\mathbf{r})}s_{\alpha\beta}\right] \quad (3.9)$$

One may see this from the theory of Green's function whose imaginary part gives the time of propagation as well as the number of states accessed in the propagation within a factor of $\hbar$. However, one may bypass all the mathematical details and assume this $\hbar$ factor difference between the two formulas and follow the subsequent discussions and analysis to see everything is consistent. The $|s_{\alpha\beta}(E)|^2$ is also easy to understand. At a given incident energy E, electrons that are incident along input channel β form an ensemble. Every member of the ensemble is not scattered to output channel α. But those that are scattered to output channel α, are done with an amplitude of $s_{\alpha\beta}$ which means for unit incident flux there are exactly $|s_{\alpha\beta}|^2$ of them. Thus, the above quantity is mathematically well defined and the physical situation is explained with Fig. 2.8 in the previous chapter. Doubling the input flux will double the number of electrons that are scattered to channel α. We cannot get linear superposition of states in the input channels as shown before in Chap. 1 Sect. 1.1. In fact, the analysis of Sect. 1.1 in Chap. 1 also naturally implies that a Slater determinant of free particle states is also not possible in the leads. Besides, such a Slater determinant or linear combination of Slater determinants in the leads will immediately result in orthogonal catastrophe which is never observed in mesoscopic systems. Numerical simulations (Mukherjee et al. 2011) suggest that Eq. 3.9 is also valid in the presence of electron-electron interaction in the sample. We get partial density of states by integrating (physically this implies averaging) $\rho_{lpd}(E, \alpha, \beta, r)$ over the spatial coordinates of the sample.

$$\rho_{pd}(E, \alpha, \beta) = -\frac{1}{4\pi i}\int_{\Omega} d^3r \left[s^*_{\alpha\beta}\frac{\delta s_{\alpha\beta}}{e\delta U(\mathbf{r})} - \frac{\delta s^*_{\alpha\beta}}{e\delta U(\mathbf{r})} s_{\alpha\beta} \right] \tag{3.10}$$

Here, Ω stands for the spatial region of the sample. To get injectivity ρ_i, we sum ρ_{lpd} in Eq. 3.9 over the outgoing channels α.

$$\rho_i(E, \beta, \mathbf{r}) = -\frac{1}{4\pi i}\sum_{\alpha} \left[s^*_{\alpha\beta}\frac{\delta s_{\alpha\beta}}{e\delta U(\mathbf{r})} - \frac{\delta s^*_{\alpha\beta}}{e\delta U(\mathbf{r})} s_{\alpha\beta} \right] \tag{3.11}$$

To get emissivity ρ_e, we sum ρ_{lpd} in Eq. 3.9 over all possible incoming channels β.

$$\rho_e(E, \alpha, \mathbf{r}) = -\frac{1}{4\pi i}\sum_{\beta} \left[s^*_{\alpha\beta}\frac{\delta s_{\alpha\beta}}{e\delta U(\mathbf{r})} - \frac{\delta s^*_{\alpha\beta}}{e\delta U(\mathbf{r})} s_{\alpha\beta} \right] \tag{3.12}$$

Injectivity and emissivity are important members of the hierarchy. Although they are not truly global parameters that mesoscopic physics talks of in the sense that they are coordinate $\mathbf{r}$ dependent, they can in principle be directly measured in an experiment and averaging over $\mathbf{r}$ is not necessary. They are local in the sense that their quantitative values will depend on the details of the sample and the exact point $\mathbf{r}$ in the sample that we want to probe. Hopefully, something like an STM tip can attach to a single point in its closest approximation. If such an STM tip delivers a current into the sample where the current can leave through one or many leads but no other lead carries a current into the sample then the current delivered will be given by injectivity. This is explained in more detail in Fig. 3.2. We have seen in Sect. 1.1 of Chap. 1 that injected current will see the DOS. What ρ_i in Eq. 3.11 means is that

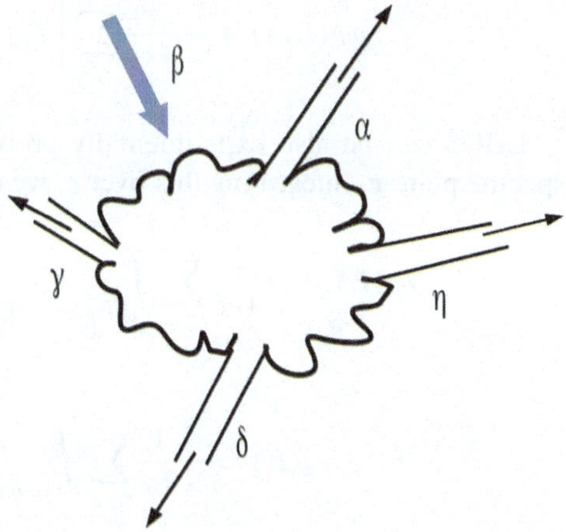

Fig. 3.2 We show a mesoscopic setup where there are many leads α, β, γ, δ, etc., attached to a sample. The lead β is special in the sense that it is an STM tip that can deliver (or draw) current to (or from) a particular point $\mathbf{r}$ in the sample. All other leads (minimum one other) draw (or deliver) current from (or to) the sample

for those particular electrons that are going from lead β (the STM tip) to α (one of the outgoing channels), the relevant part of DOS that determine the injected current is ρ_i. Equation 3.10 involves an integration over r and so if we sum Eq. 3.10 over β, then we get emittance of a lead α and if we sum Eq. 3.10 over α then we get injectance of lead β. If there are N leads attached to a sample such that only one lead indexed β delivers current to the sample (averaged over $\mathbf{r}$) and all other leads carry current away from the sample then the injectance gives the current delivering capacity of this lead α. Thus, current delivered at a point $\mathbf{r}$ (injectivity) is a much simpler concept rather than injectance because experimentally sweeping an STM tip over all $\mathbf{r}$ values (to average over $\mathbf{r}$) do not appear practical and theoretically trying to determine functional derivatives with respect to the local potential is impossible as then we need to know the exact functional form of the potential inside the sample. For large samples, this problem can be bypassed through ensemble averaging which we are not allowed to do in case of mesoscopic samples. Since we will provide a solution to this problem in Chap. 5, let us anyway proceed with this approach and treat all members of the hierarchy as physical. Injectance can thus be defined as

$$\rho(\beta, E) = \int_{\Omega} d^3r \sum_{\alpha} -\frac{1}{4\pi i} \left[s^*_{\alpha\beta} \frac{\delta s_{\alpha\beta}}{e\delta U(\mathbf{r})} - \frac{\delta s^*_{\alpha\beta}}{e\delta U(\mathbf{r})} s_{\alpha\beta} \right] \tag{3.13}$$

Just like resistance characterizes the dissipative capacity of a sample although strongly dependent on the nature of leads, injectance characterizes the current delivering capacity of a lead that is strongly dependent on the details of the sample. Local density of states (LDOS) can be defined by summing the RHS of Eq. 3.9 over α and β.

$$\rho_{ld}(E, \mathbf{r}) = -\frac{1}{4\pi i} \sum_{\alpha\beta} \left[s^*_{\alpha\beta} \frac{\delta s_{\alpha\beta}}{e\delta U(\mathbf{r})} - \frac{\delta s^*_{\alpha\beta}}{e\delta U(\mathbf{r})} s_{\alpha\beta} \right] \tag{3.14}$$

LDOS can be also experimentally probed directly with an STM attached to a specific point $\mathbf{r}$. Integrating this over r, we obtain

$$\rho_d(E) = -\frac{1}{4\pi i} \sum_{\alpha\beta} \int_{sample} dr \left[s^*_{\alpha\beta} \frac{\delta s_{\alpha\beta}}{e\delta U(\mathbf{r})} - \frac{\delta s^*_{\alpha\beta}}{e\delta U(\mathbf{r})} s_{\alpha\beta} \right] \tag{3.15}$$

$$\rho_d(E) = -\frac{1}{2\pi} \sum_{\alpha\beta} \int_{sample} dr \left[|s_{\alpha\beta}|^2 \frac{\delta\theta_{s_{\alpha\beta}}}{e\delta U(\mathbf{r})} \right] \tag{3.16}$$

This is the mesoscopic version of Friedel sum rule that relates scattering phase shift to DOS. In this equation too, we get the derivative of the scattering phase shift with respect to the local potential $\mathbf{r}$ along with an integration over $\mathbf{r}$ and so this is not of much use to an experimentalist. One may consider the following substitution

$$\int_{global} dr \frac{\delta}{e\delta U(\mathbf{r})} = -\frac{d}{dE} \tag{3.17}$$

$$\rho_d(E) = \frac{1}{2\pi} \sum_{\alpha\beta} \left[|s_{\alpha\beta}|^2 \frac{d\theta_{s_{\alpha\beta}}}{dE} \right] \tag{3.18}$$

This is the FSR that appears in text books. This is true because the global integration on LHS implies that the potential can be increased (or decreased) by a constant amount globally which is identical to not shifting the potential but decreasing (or increasing) the incident energy. An experimentalist can easily achieve this by changing the Fermi energy but what appears in Eq. 3.16 is not a global integration but a sample integration. For a large bulk system, one need not make a difference between the two but for a mesoscopic system we may only write

$$\int_{sample} dr \frac{\delta}{e\delta U(\mathbf{r})} \approx -\frac{d}{dE} \tag{3.19}$$

Thus,

$$\rho_d(E) \approx \frac{1}{2\pi} \left[|s_{\alpha\beta}|^2 \frac{d\theta_{s_{\alpha\beta}}}{dE} \right] \tag{3.20}$$

Thus, using the analogy with Sect. 1.1 of Chap. 1 where we show that current is nev and differential current is $\frac{dn}{dE}evdE$, one obtains (e or electronic charge can be set to unity and, if properly normalized wavefunctions are taken, then we can also drop the v factor).

$$\rho(\beta, E) = \int_{sample} dxdy \sum_{k_\beta} |\psi(x, y, \beta)|^2 \delta(E - E_{\beta,k_\beta}) = \tag{3.21}$$

$$\int_\Omega d^3r \sum_\alpha -\frac{1}{4\pi i}\left[s^*_{\alpha\beta}\frac{\delta s_{\alpha\beta}}{e\delta U(\mathbf{r})} - \frac{\delta s^*_{\alpha\beta}}{e\delta U(\mathbf{r})}s_{\alpha\beta}\right] \tag{3.22}$$

In 1D, it was shown by Leavens and Aers (1989) that the correction term can be determined in 1D to give

$$\rho(1, E) = \frac{1}{2\pi}\left[|r|^2\frac{d\theta_r}{dE} + |t|^2\frac{d\theta_t}{dE} + \frac{m_e|r|}{\hbar k^2}sin(\theta_r)\right] \tag{3.23}$$

which means the last term is the correction term due to interference effects. So that if we consider a semi-classical regime where the electron behaves like a classical particle, then $sin(\theta_r)$ goes to zero and we recover the text book FSR (in fact in practice ensemble averaging washes out any dependence on $sin(\theta_r)$ and make large systems effectively behave semi-classically). We will show that when phase drops occur $\sin(\theta_r)$ can become zero in a quantum regime making Eq. 3.20 exact in a quantum regime implying something else is going on. And that will also imply a practical solution to study mesoscopic systems by bypassing sweeping STM tips or doing functional derivatives. These will be carefully explained in the next few chapters.

References

1. M. Buttiker, *Time in Quantum Mechanics* (Springer, 2001), pp. 279–303
2. C.R. Leavens, G.C. Aers, Phys. Rev. B **39**, 1202 (1989)
3. S. Mukherjee et al., Physica E **44**, 62 (2011)
4. J.M. Ziman, *Principles of the Theory of Solids*, 2nd edn. (Cambridge University Press, 1979)

Chapter 4
Scattering in Q1D

In this chapter, we briefly outline how to solve a scattering problem for a realistic mesoscopic system giving the references to where the details can be found. Our motivation is to present some extra calculations along the same line and point out some new results for these particular systems which will be generalized in the next two chapters from general principles and theorems. These extra calculations were first reported in Singha Deo (2007), Satpathy and Singha Deo (2012). We will point out that from these solutions one can verify the phase slips of π that has somehow eluded attention before our works. Its implications will be discussed too. These scattering problems are generally hard to solve and numerically even simple cases require a good amount of computation time (Kalman et al. 2008). The only potential that can be solved exactly is that of a delta function potential (Bagwell 1990) to give the scattering amplitudes. We will briefly outline his solution and then proceed to give a detailed derivation of the DOS for the same which is very instructive.

Before going into the mathematical analysis, we discuss how the Fano resonances and the associated phase slips appear very naturally in quantum wires. Consider a simplified version of Fig. 2.8 as shown in Fig. 4.1. Here, the shaded region of Fig 2.8 has been simplified into a rectangular block extending from $x = -a$ to $x = a$ and from $y = -c/2$ to $y = c/2$. The leads extend from $y = -b/2$ to $y = b/2$. The mathematical solution for this system will be given in Sect. 4.1. This system can be further simplified by making c equal to b in which case we will get the structure shown in Fig. 4.2c. All the systems in Figs. 2.8, 4.1 and 4.2c will exhibit Fano resonances along with discontinuous phase shifts of π of the nature discussed in Sect. 2.2. The physical reasons can be diagrammatically explained as follows. For example, consider the modes in the leads in Fig. 2.8. The dotted sinusoidal curve is the wavefunction for the ground state in an infinite potential well as the potential in the direction transverse to the direction of current flow is taken as hard wall. This is just the wavefunction of the first state in Fig. 2.1, but now turned vertically. The wavefunction in the propagating direction will be plane waves with usual parabolic dispersion curve as we will show in Sect. 4.1. The dotted parabolic curve in Fig. 4.2d is showing this dispersion (it is basically a plot of energy versus wavenumber). The

P. Singha Deo, *Mesoscopic Route to Time Travel*,
https://doi.org/10.1007/978-981-16-4465-8_4

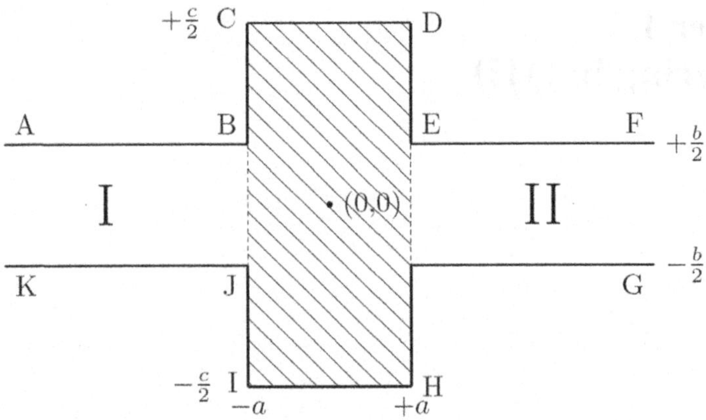

Fig. 4.1 A simplified version of the system in Fig. 2.8

solid sinusoidal curve in the leads of Fig. 2.8 is the first excited state (depicted as second state in Fig. 2.1) in an infinite potential well and, for this, we will get another parabolic dispersion curve shown by dashed curve in Fig. 4.2d. This parabola will be at a higher energy relative to the dotted parabola as the first excited state in the transverse direction has higher energy than the ground state (will be shown mathematically in Sect. 4.1). Similarly, corresponding to the dashed sinusoidal curve in Fig. 2.8, we will get the solid parabola in Fig. 4.2d. These parabolic dispersion curves correspond to scattering states only. If potential V in the sample region is negative (say $-V_0$) then corresponding to each parabola there will be bound states as well whose energies will be below the threshold energy of the parabolas. Thus, the solid parabola will have some bound states shown by the solid straight lines below the solid parabola in Fig. 4.2d. The dashed parabola will also have some bound states shown by the dashed straight lines below the dashed parabola in Fig. 4.2d. Now the dashed bound states will not be true bound states as they are degenerate with a scattering state (shown by an arrow), of the dotted channel. We know that degeneracy between a bound state and a scattering state will always produce a Fano resonance (Porod et al. 1993) that will be accompanied by a discontinuous phase change of π at the zeroes. However, the mesoscopic picture is rather different for the reservoirs as we will see in subsequent chapters. There will be no true transmission zero but a minimum in the transmission and no discontinuous phase change but instead a sharp and continuous phase drop that will dramatically change our idea about Friedel sum rule like formulas and our understanding of time. Phase drop is a rather novel phenomenon because our intuitive understanding of scattering says that time delays in electron propagation and phase accumulation of the wavefunction are related. So phase drop may naively suggest time gains rather than delays. But one has to carefully isolate what is a true physical phenomenon and what is artifact of approximations and this will be done in subsequent chapters.

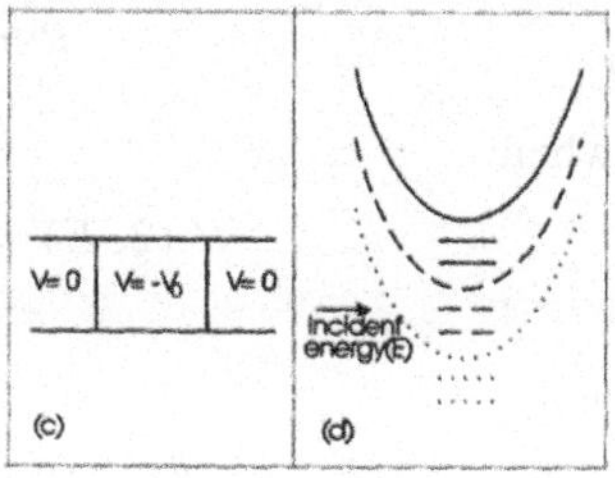

Fig. 4.2 An intuitive picture of getting Fano resonances in mesoscopic systems

For two transverse channels as discussed earlier, the semi-classical injectance of k_1 channel of left lead is given by

$$\rho(1,E) \approx \frac{1}{2\pi}\left[\mid r_{11} \mid^2 \frac{d\theta_{r_{11}}}{dE} + \mid r_{21} \mid^2 \frac{d\theta_{r_{21}}}{dE} + \mid t_{11} \mid^2 \frac{d\theta_{t_{11}}}{dE} + \mid t_{21} \mid^2 \frac{d\theta_{t_{21}}}{dE} \right] \quad (4.1)$$

We can break this up as

$$\rho(1,E) \approx \rho^L(1,E) + \rho^R(1,E) \quad (4.2)$$

where

$$\rho^L(1,E) = \frac{1}{2\pi}\left[\mid r_{11} \mid^2 \frac{d\theta_{r_{11}}}{dE} + \mid r_{21} \mid^2 \frac{d\theta_{r_{21}}}{dE} \right] \quad (4.3)$$

and

$$\rho^R(1,E) = \frac{1}{2\pi}\left[\mid t_{11} \mid^2 \frac{d\theta_{t_{11}}}{dE} + \mid t_{21} \mid^2 \frac{d\theta_{t_{21}}}{dE} \right] \quad (4.4)$$

That is, $\rho^L(1,E)$ consist of reflection channels and $\rho^R(1,E)$ consist of transmission channels. The correction term depends on parameters of the incident channel only and gives the following identity:

$$\rho(1,E) = \rho^L(1,E) + \rho^R(1,E) + \frac{1}{2\pi}\frac{m_e \mid r_{11} \mid}{\hbar k_1^2} sin(\theta_{r_{11}}) \quad (4.5)$$

While this was known in 1D (Eq. 3.23) earlier, the explicit calculation here reveal that this is also true for the multichannel situation and the correction term depend solely on the input channel parameters. Similarly, semi-classical injectance of k_2 channel is,

$$\rho(2,E) \approx \frac{1}{2\pi}\left[\mid r_{22} \mid^2 \frac{d\theta_{r_{22}}}{dE} + \mid r_{12} \mid^2 \frac{d\theta_{r_{12}}}{dE} + \mid t_{22} \mid^2 \frac{d\theta_{t_{22}}}{dE} + \mid t_{12} \mid^2 \frac{d\theta_{t_{12}}}{dE} \right] \quad (4.6)$$

We can break this up as

$$\rho(2, E) \approx \rho^L(2, E) + \rho^R(2, E) \tag{4.7}$$

where

$$\rho^L(2, E) = \frac{1}{2\pi}\left[\mid r_{22} \mid^2 \frac{d\theta_{r_{22}}}{dE} + \mid r_{12} \mid^2 \frac{d\theta_{r_{12}}}{dE} \right] \tag{4.8}$$

and

$$\rho^R(2, E) = \frac{1}{2\pi}\left[\mid t_{22} \mid^2 \frac{d\theta_{t_{22}}}{dE} + \mid t_{12} \mid^2 \frac{d\theta_{t_{12}}}{dE} \right] \tag{4.9}$$

and with the correction term we get

$$\rho(2, E) = \rho^L(2, E) + \rho^R(2, E) + \frac{1}{2\pi} \frac{m_e \mid r_{22} \mid}{\hbar k_2^2} sin(\theta_{r_{22}}) \tag{4.10}$$

The RHS of Eqs. 4.2 and 4.7 can be determined from the S-matrix. This S-matrix approach is easily accessible to experimentalists. But the correction term can be very large in mesoscopic systems that are in the quantum regime. The physical interpretation of the correction term also needs a major change.

4.1 A Typical Scattering Problem in Q1D

An analytic treatment of scattering in Q1D for a general system of the type shown in Fig. 2.8 is given below in terms of partial wave analysis. The purpose is to show that discontinuous π phase shifts occur in such a realistic system for the reasons mentioned in the first two paragraphs of this chapter. The analysis can be adopted to solve the system in Fig. 4.1 explicitly and has been done earlier (Singha Deo et al. 1998). Consider a quasi one dimensional (Q1D) system with scattering potential $V_g(x, y)$ shown in Fig. 2.8 by the shaded region. Left reservoir act as source of electrons and right reservoir as sink. As usual the x-direction is along the source to sink direction and the y-direction is perpendicular to it. The confinement potential in the leads, i.e., in regions L and R, is taken to be hard wall in the y-direction (or, transverse direction), given by

$$V(x, y) = V(y) = \infty \text{ for } \mid y \mid \geq \frac{b}{2} \text{ and } \mid x \mid \geq a$$

$$= 0 \text{ for } \mid y \mid < \frac{b}{2} \text{ and } \mid x \mid \geq a \tag{4.11}$$

The potential in the shaded region is of the form

$$V(x, y) = \infty \text{ for } \mid y \mid \geq \frac{c}{2} \text{ and } \mid x \mid < a$$

$$= V_g(x, y) \text{ for } | y |< \frac{c}{2} \text{ and } | x |< a \tag{4.12}$$

Thus we have taken the system to be embedded in 2D and the reason for this will soon be clear. The Schrödinger equation in two dimensions is,

$$\left[-\frac{\hbar^2}{2m_e}\left(\frac{\partial^2}{\partial x^2}+\frac{\partial^2}{\partial y^2}\right)+V(x, y)\right]\Psi(x, y) = E\Psi(x, y) \tag{4.13}$$

where m_e is the effective mass of the electron. Let us first solve for the scattering states or the asymptotic states in the leads, that is in regions I and II. There the Schrödinger equation reduces to

$$\left[-\frac{\hbar^2}{2m_e}\left(\frac{\partial^2}{\partial x^2}+\frac{\partial^2}{\partial y^2}\right)+V(y)\right]\Psi(x, y) = E\Psi(x, y) \tag{4.14}$$

Here the x and y components decouple so that the we can write

$$\Psi(x, y) = c_m(x)\chi_m(y) \tag{4.15}$$

Thus the y-component is,

$$\left[-\frac{\hbar^2}{2m_e}\frac{d^2}{dy^2}+V(y)\right]\chi_m(y) = \varepsilon_m\chi_m(y) \tag{4.16}$$

And the x-component is,

$$-\frac{\hbar^2}{2m_e}\frac{d^2}{dx^2}c_m(x) = (E-\varepsilon_m)c_m(x) \tag{4.17}$$

Note that $V(y)$ is just an infinite potential in 1D y direction, just as the one in Fig. 2.1 in x direction. Thus we get from Eq. 4.16

$$\chi_m(y) = \sqrt{\frac{2}{b}}\sin\frac{m\pi}{b}(y+\frac{b}{2}) \tag{4.18}$$

and

$$\varepsilon_m = \frac{\hbar^2 m^2\pi^2}{2\, m_e b^2}. \tag{4.19}$$

E is the energy of incidence from reservoir L. From Eq. 4.17 we get

$$c_m(x) = e^{\pm ik_m x} \tag{4.20}$$

and

$$(E - \varepsilon_m) = \frac{\hbar^2 k_m^2}{2m_e} \tag{4.21}$$

Thus we can write from Eqs. 4.19 and 4.21

$$E = E_{m,k_m} = \frac{\hbar^2 m^2 \pi^2}{2m_e b^2} + \frac{\hbar^2 k_m^2}{2m_e} \tag{4.22}$$

Note that if we had considered the z-direction in Eq. 4.13 then this Eq. 4.22 would have been

$$E = E_{m,n,k_{mn}} = \frac{\hbar^2 m^2 \pi^2}{2m_e b^2} + \frac{\hbar^2 q^2 \pi^2}{2m_e d^2} + \frac{\hbar^2 k_m^2}{2m_e} \tag{4.23}$$

where d is the finite width of the quantum wire in the z-direction. The way these samples are made we get $d << b$ which means only q=1 state is populated as the rest are very high in energy. So we can set the constant energy of the only mode in the z-direction to zero by choice of the reference frame of energy.

When the potential have symmetry in x-direction, i.e., $V_g(x, y) = V_g(-x, y)$, it is suitable to write solutions as given below (Bayman and Mehoke 1983).

$$c_m(x) = \frac{c_n^e(x) - c_m^o(x)}{2} \tag{4.24}$$

where, $c_m^e(x) = c_n^e(-x)$ and $c_m^o(x) = -c_m^o(-x)$. Now from Eq. 4.20 one can generally write

$$c_n^e(x) = \sum_{m=1}^{\infty} \frac{1}{\sqrt{k_m}} (\delta_{mn} e^{-ik_m x} - S_{mn}^e e^{ik_m x}) \tag{4.25}$$

$$c_n^o(x) = \sum_{m=1}^{\infty} \frac{1}{\sqrt{k_m}} (\delta_{mn} e^{-ik_m x} - S_{mn}^o e^{ik_m x}) \tag{4.26}$$

This defines from Eqs. 4.15, 4.18, 4.25, and 4.26

$$\Psi_n^e(x, y) = \sum_{m=1}^{\infty} \frac{\sqrt{2}}{\sqrt{k_m b}} (\delta_{mn} e^{-ik_m x} - S_{mn}^e e^{ik_m x}) \sin \frac{m\pi}{b}(y + \frac{b}{2}) \tag{4.27}$$

$$\Psi_n^o(x, y) = \sum_{m=1}^{\infty} \frac{\sqrt{2}}{\sqrt{k_m b}} (\delta_{mn} e^{-ik_m x} - S_{mn}^o e^{ik_m x}) \sin \frac{m\pi}{b}(y + \frac{b}{2}) \tag{4.28}$$

Then both transmitted wavefunction (at $x > a$) and reflected wavefunction (at $x < -a$) are given by Eqs. 4.27 and 4.28 as it is defined for $|x| \geq a$.

For any arbitrary potential one can use the tight binding approximation to solve a problem but this is a drastic approximation that cannot be applied without a proper scrutiny. It reduces a problem connected to the continuum to that of solving a finite dimensional matrix equation. Till date there is no study or effort to account for the evanescent modes that play a very crucial role in mesoscopic systems. Below we show how these modes appear very naturally and play a crucial role in determining mesoscopic properties. E in Eq. 4.22 can be adjusted by adjusting the Fermi energy of left reservoir, that

$$\frac{\pi^2\hbar^2}{2m_e b^2} < E < \frac{4\pi^2\hbar^2}{2m_e b^2} \tag{4.29}$$

In this regime k_1 is real. For $m > 1$, energy conservation in Eq. 4.22 is not violated as k_m^2 can become negative. That yields evanescent solutions for $m > 1$ with $k_m \to i\kappa_m$. One can also choose energy in the range

$$\frac{4\pi^2\hbar^2}{2m_e b^2} < E < \frac{9\pi^2\hbar^2}{2m_e b^2} \tag{4.30}$$

when k_1 and k_2 are real and all other k_3, k_4, k_5 etc., are imaginary as can be seen from Eq. 4.22. Inclusion of the evanescent modes is very important to get the correct solutions (quantitative as well as qualitative) as we show below.

Any function can be written as a sum of an even function and odd function and any square matrix can be written as a sum of a symmetric matrix and an anti-symmetric matrix, and so the wavefunction in the scattering region ($-a < x < a$, where the potential is given in Eq. 4.12) can be written as

$$\Psi(x, y) = \frac{\eta_m^e(x, y) + \eta_m^o(x, y)}{2} \quad for \quad -a < x < a \tag{4.31}$$

where

$$\eta_m^e(x, y) = \sum_{n=1}^{\infty} d_n \zeta_n^e(x, y) \tag{4.32}$$

$$\eta_m^o(x, y) = \sum_{n=1}^{\infty} d_n \zeta_n^o(x, y) \tag{4.33}$$

ζ_n^e and ζ_n^o are the basis states that satisfy the condition that they go to zero at the upper edge and lower edge of the shaded region in Fig. 2.8. The simplified upper edge in Fig. 4.1 is BCDE and for this also the general forms in Eqs. 4.31, 4.32 and 4.33 are valid. One can define the matrix elements,

$$F_{m,n}^{eo} = \frac{\sqrt{2}}{\sqrt{b}(k_m k_n)^{\frac{1}{2}}} \int_{-b/2}^{b/2} \chi_m(y) \left(\frac{\partial \zeta_n^{eo}}{\partial x}\right)_{x=a} dy \tag{4.34}$$

Here eo stands for even or odd, i.e., e/o. One can match the wavefunction and conserve the current at $x = \pm a$ for all y between $-b/2$ and $b/2$ to obtain a matrix equation given by,

$$\sum_{n=1}^{\infty}\left[F^{eo}_{rn} - i\delta_{rn}\right] e^{ik_n a} S^{eo}_{nm} = \left[F^{eo}_{rm} + i\delta_{rm}\right] e^{-ik_m a} \tag{4.35}$$

For the energy interval given in Eq. 4.29, solving for S^{eo}_{mn}, we can find the scattering matrix elements $\tilde{r}_{11}$, $\tilde{t}_{11}$, etc. where

$$\tilde{r}_{11} = -\frac{(S^{o}_{11} + S^{e}_{11})}{2} \tag{4.36}$$

$$\tilde{t}_{11} = \frac{(S^{o}_{11} - S^{e}_{11})}{2} \tag{4.37}$$

Then k_1 is real and from Eqs. 4.24, 4.25, 4.26, 4.36 and 4.37

$$\begin{aligned} c_1(x) &= e^{ik_1 x} + \tilde{r}_{11} e^{-ik_1 x} \text{ for} x < -a \\ &= \tilde{t}_{11} e^{ik_1 x} \text{ for } x > a \end{aligned}$$

As explained before when we have only one propagating channel in the two leads one can have true bound states below the threshold energy of propagation. Bound states can be determined from the singularities of the matrix Eq. (4.35). That is

$$det\left[F^{eo}_{cc} - i1\right] = 0 \tag{4.38}$$

Here 'cc' means evanescent channel (or closed channel) for which both k_m and k_n in Eq. 4.35 are imaginary. For the situation wherein E can be so adjusted that

$$\frac{\pi^2\hbar^2}{2\,m_e b^2} < E < \frac{4\pi^2\hbar^2}{2\,m_e b^2} \tag{4.39}$$

we have argued from Fig. 4.2 that there will be a degeneracy between a scattering state and a bound state resulting in a Fano resonance. In this regime

$$S^{eo}_{11} = e^{-2ik_1 a}\frac{G^{eo} + i}{G^{eo} - i} = e^{2i arccot[G^{eo}]} = e^{2i\theta^{eo}} \tag{4.40}$$

where,

$$G^{eo} = F^{eo}_{11} - \sum_{m=2,n=2} F^{eo}_{1n}\left[\left(F^{eo}_{cc} - i1\right)^{-1}\right]_{nm} F^{eo}_{m1} \tag{4.41}$$

and

$$\theta^{eo} = arccot[G^{eo}] \tag{4.42}$$

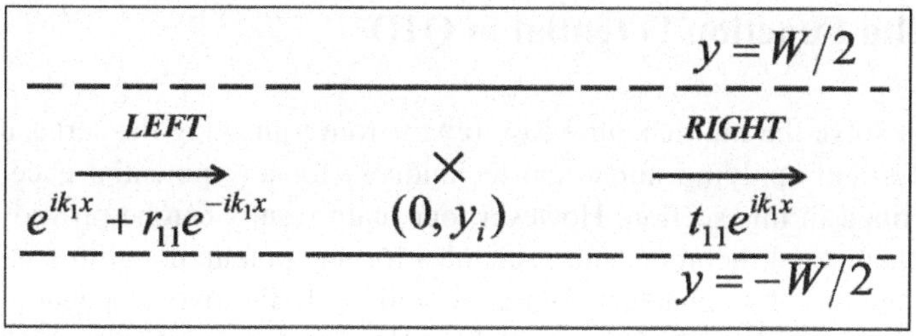

Fig. 4.3 A Dirac delta function potential in a quantum wire

See (Bayman and Mehoke 1983) for details. Here scattering phase shift θ^{eo} is defined with respect to phase shift in the absence of scatterer. We define new variables,

$$\phi = \theta^e - \theta^o \text{ and } \theta_r = \theta^e + \theta^o \tag{4.43}$$

we get transmission amplitude and reflection amplitude

$$\tilde{t}_{11} = i\sin(\phi)e^{i\theta_r} \text{ and } \tilde{r}_{11} = -\cos(\phi)e^{i\theta_r} \tag{4.44}$$

The correction term of Eq. 4.5 in case of this system is $\frac{m_e|\tilde{r}_{11}|}{\hbar k_1^2}\sin(\theta_r)$.

Threshold energy E for the second channel is given by $E = \frac{4\pi^2\hbar^2}{2m_e b^2}$ (i.e., above this energy k_2 becomes real and hence propagating). Below this energy the second channel can have bound states at energies E that satisfy Eq. 4.38. At these energies the first channel is propagating. S_{11} is given by Eq. 4.40. But at bound state energy $det\left[F^{eo}_{cc} - i1\right] = 0$, and so G^{eo} will diverge as it includes matrix elements of $\left[F^{eo}_{cc} - i1\right]^{-1}$ as can be seen from Eq. 4.41. In other words for resonant condition (Eq. 4.38), G_{eo} diverges (see Eq. 4.41) and so from Eq. 4.42

$$\theta^e = p\pi \text{ and } \theta^o = q\pi \tag{4.45}$$

where p and q are integers. Therefore, from Eq. 4.43,

$$\sin(\theta_r) = \sin(\theta^e + \theta^o) = 0 \tag{4.46}$$

Thus, we have shown that the correction term in Eq. 4.5 is zero because $\sin(\theta_r) = 0$. At this point $|r| = 1$ as can be seen from Eqs. 4.43 and 4.44. As stated earlier, if the correction term can be ignored or set to zero then the semi-classical formulae can be very useful to experimentalists.

4.2 Delta Function Potential in Q1D

We cannot solve the multichannel case (apart from truncating the series of evanescent modes and applying numerical techniques) for any potential except for the case described in this section. However, our main results can be proved generally using mathematical theorems and principles for any potential and this will be presented in the next two chapters. Here, we will study the two propagating channel (with arbitrary number of evanescent channel) case for a particular potential, i.e., $V_g(x, y) = \gamma\delta(x)\delta(y - y_j)$, because it can be exactly solved and certain facts can be exhibited explicitly. In Fig. 4.3, the shaded region of Fig. 2.8 reduces to give a two-dimensional (2D) quantum wire with a delta function potential at position $(0, y_j)$ marked X. The dotted lines in Fig. 4.3 represent the fact that the quantum wire is connected to electron reservoirs.

We will first calculate this injectance from internal wavefunction $\psi(x, y, 1)$ which allows one to make numerical verification of the semi-classical formulae in Eqs. 4.1, 4.6 and their correction terms in Eqs. 4.5 and 4.10. We know that when incidence is along n=1 channel

$$\rho(1, E) = \int_{-\infty}^{\infty} dx \int_{-\frac{W}{2}}^{\frac{W}{2}} dy \sum_{k_1} \mid \psi(x, y, 1) \mid^2 \delta(E - E_{1,k_1}) \tag{4.47}$$

Here, E_{1,k_1} is the energy of the incident electron along channel $n = 1$ with wavevector $k_{n=1}$ or k_1. Summing over n, one can get the standard expression for DOS according to spectral theorem. Note that, although k_1 can vary continuously, one can write sum over k_1 precisely because of the spectral theorem. One can write for the system represented in Fig. 4.2 (Bagwell 1990) when incidence is along n=1 channel

$$\psi(x, y, 1) = \sum_m f_m(x, 1)\chi_m(y) \tag{4.48}$$

Here, $\chi_m(y)s$ are solutions in Eq. 4.18 in the leads in the transverse direction which is an infinite square well potential in y-direction. $\chi_m(y)s$ form a complete set and Eq. 4.48 is derived from the fact that at a given point x, the wavefunction in the scattering region can be expanded in terms of $\chi_m(y)s$. We will present our results for the case of two propagating channels, but the analysis and results are same for any number of channels. $f_m s$ are generally of the form given below

$$f_1(x, 1) = e^{ik_1x} + r_{11}e^{-ik_1x} \quad for \quad x < 0 \tag{4.49}$$

$$= t_{11}e^{ik_1x} \quad for \quad x > 0 \tag{4.50}$$

$$f_2(x, 1) = r_{21}e^{-ik_2x} \quad for \quad x < 0 \tag{4.51}$$

$$= t_{21}e^{ik_2x} \quad for \quad x > 0 \tag{4.52}$$

and for $m > 2$,

$$f_m(x, 1) = r'_{m1} e^{\kappa_m x} \quad for \;\; x < 0 \tag{4.53}$$

$$= t'_{m1} e^{-\kappa_m x} \quad for \;\; x > 0 \tag{4.54}$$

Here, r_{mn}, t_{mn} r'_{mn} and t'_{mn} are unknowns to be determined. The scattering problem described above can be solved using mode matching technique. The reflection amplitudes are given by Bagwell (1990),

$$r_{mn}(E) = -\frac{i\frac{\Gamma_{mn}}{2\sqrt{k_m k_n}}}{1 + \sum_e \frac{\Gamma_{ee}}{2\kappa_e} + i\sum_m \frac{\Gamma_{mm}}{2k_m}} \tag{4.55}$$

Γ_{mn} is the coupling strength between mth and nth modes, given by

$$\Gamma_{mn} = \frac{2m\gamma}{h^2} sin\left[\frac{m\pi}{W}\left(y_j + \frac{W}{2}\right)\right] sin\left[\frac{n\pi}{w}\left(y_j + \frac{W}{2}\right)\right] \tag{4.56}$$

The transmission amplitudes are given by

$$t_{mn} = r_{mn}(E) \tag{4.57}$$

for $m \neq n$ and

$$t_{mm}(E) = 1 + r_{mm}(E) \tag{4.58}$$

Along the same line, one can show that

$$r'_{m1} = t'_{m1}(E) = -\frac{\frac{\Gamma_{m1}}{2\kappa_m}}{1 + \sum_e \frac{\Gamma_{ee}}{2\kappa_e} + i\left(\frac{\Gamma_{11}}{2k_1} + \frac{\Gamma_{22}}{2k_2}\right)} \tag{4.59}$$

$\sum_e$ denote sum over evanescent modes and run from 3 to ∞, while $\sum_m$ denote the same for propagating modes (i.e., $m = 1$ and $m = 2$).

If the delta function potential is negative ($\gamma < 0$), then there can be a bound state as argued before. As explained in Fig. 4.2d, it will always be a quasi-bound state. The quasi-bound state for channel $m = s$ is given by (Bagwell 1990),

$$1 + \sum_{e=s}^{\infty} \frac{\Gamma_{ee}}{2k_e} = 0 \tag{4.60}$$

Only bound state for $m = 1$ channel is a true bound state and it is given by the solution to the following equation: $1 + \sum_{e=1}^{\infty} \frac{\Gamma_{ee}}{2k_e} = 0$. The bound state for $m = 2$ is given by

$$1+\sum_{e=2}^{\infty}\frac{\Gamma_{ee}}{2k_e}=0 \tag{4.61}$$

At this energy, we get a bound state for $m = 2$, but at that very energy, $m = 1$ channel (or k_1 channel) is a propagating channel as can be seen in Eq. 4.29. Hence, the bound state given by this Eq. 4.61 is a quasi-bound state or a resonance. The delta function in Eq. 4.47 summed over k_1 essentially yield a factor $\frac{1}{hv_1}$, where $v_1 = \frac{\hbar k_1}{m}$. Substituting for $\psi()$ from Eq. 4.48 and using the orthogonality of $\chi_m(y)s$, we get

$$\rho(1,E)=\frac{1}{hv_1}\left[\int_{-\infty}^{\infty}dx\sum_m \mid f_m(x,1)\mid^2\right] \tag{4.62}$$

$$=\frac{1}{hv_1}\left[\int_{-\infty}^{\infty}dx\mid f_1(x,1)\mid^2+\int_{-\infty}^{\infty}\mid f_2(x,1)\mid^2+\int_{-\infty}^{\infty}dx\mid f_3(x,1)\mid^2+\int_{-\infty}^{\infty}dx\mid f_4(x,1)\mid^2+\cdots\right] \tag{4.63}$$

Substituting the values of $f_m(x, 1)s$ from Eqs. 4.49–4.52, we get

$$\rho(1,E)=\frac{1}{hv_1}\left[\int_{-\infty}^{0}dx[1+\mid r_{11}\mid^2+2\mid r_{11}\mid cos(2k_1x+\phi_1)]\right.$$

$$\left.+\int_{0}^{\infty}dx\mid t_{11}\mid^2+\int_{-\infty}^{0}dx\mid r_{21}\mid^2+\int_{0}^{\infty}dx\mid t_{21}\mid^2+H\right. \tag{4.64}$$

Here, $r_{11} =\mid r_{11}\mid e^{-i\phi_1}$. Note that, for $m > 2$,

$$H=\int_{-\infty}^{\infty}dx\sum_{m>2}\mid f_m(x,1)\mid^2=\sum_{m>2}\mid t'_{m1}\mid^2\left[\int_{-\infty}^{0}e^{2\kappa_m x}dx+\int_{0}^{\infty}e^{-2\kappa_m x}dx\right]$$

$$=\sum_{m>2}\frac{\mid t'_{m1}\mid^2}{\kappa_m} \tag{4.65}$$

Thus,

$$\rho(1,E)=\frac{1+\mid r_{11}\mid^2+\mid r_{21}\mid^2}{hv_1}\int_{-\infty}^{0}dx+\frac{\mid t_{11}\mid^2+\mid t_{21}\mid^2}{hv_1}\int_{0}^{\infty}dx$$

$$+\frac{2\mid r_{11}\mid}{hv_1}\int_{-\infty}^{0}dx cos(2k_1x+\phi_1)$$

$$+\frac{1}{hv_1}\left(\frac{\mid t'_{31}\mid^2}{\kappa_3}+\frac{\mid t'_{41}\mid^2}{\kappa_4}+\cdots\right) \tag{4.66}$$

Adding and subtracting the following terms,

$$\frac{|t_{11}|^2}{hv_1}\int_{-\infty}^{0} dx, \quad \frac{|t_{21}|^2}{hv_1}\int_{-\infty}^{0} dx,$$

and using the fact that,

$$\frac{|t_{11}|^2}{hv_1}\int_{-\infty}^{0} dx = \frac{|t_{11}|^2}{hv_1}\int_{0}^{\infty} dx$$

and $|r_{11}|^2 + |r_{21}|^2 + |t_{11}|^2 + |t_{21}|^2 = 1$, we get,

$$\rho(1, E) = \frac{1}{hv_1}\int_{-\infty}^{\infty} dx + \frac{|r_{11}|}{hv_1}\int_{-\infty}^{\infty} dx cos(2k_1 x + \phi_1)$$
$$+ \frac{1}{hv_1}\left(\frac{|t_{31}|^2}{\kappa_3} + \frac{|t_{41}|^2}{\kappa_4} + \cdots\right) \tag{4.67}$$

$\frac{1}{hv_1}\int_{-\infty}^{\infty} dx = P_0(E)$ is the injectance in the absence of scatterer. Again if the scattering phase shift is defined with respect to the phase shift in absence of scatterer then this term can be dropped as shown in Eqs. 1.95 and 1.97. Now,

$$\int_{-\infty}^{\infty} dx cos(2k_1 x + \phi_1) = \delta(2k_1) cos(\phi_1)$$

As $k_1 = 0$ is not a propagating state contributing to transport, this term in Eq. 4.67 becomes zero, and we are left with,

$$\rho(1, E) = \frac{1}{hv_1}\left(\frac{|t'_{31}|^2}{\kappa_3} + \frac{|t'_{41}|^2}{\kappa_4} + \cdots\right) \tag{4.68}$$

Similarly, for incidence along channel 2, one can obtain the injectance.

$$\rho(2, E) = \frac{1}{hv_2}\left(\frac{|t'_{32}|^2}{\kappa_3} + \frac{|t'_{42}|^2}{\kappa_4} + \cdots\right) \tag{4.69}$$

In Fig. 4.4, we make a comparison between the semi-classical injectance $\rho(1, E)$ (solid curve) from Eq. 4.1 and the same from the exact formula in Eq. 4.68 (dash-dot curve). The resonance condition in Eq. 4.60 is satisfied at $EW = 84.29$ at which the injectance peaks. Surprisingly, at the resonance, the semi-classical formula becomes exact. The same happens for $\rho(2, E)$ for which the semi-classical formula in Eq. 4.6 gives the dotted curve and the exact formula in Eq. 4.69 gives the dashed curve. That this is not a special situation for the delta function potential but a general feature of Fano resonances will be shown in the next chapter. We have also stated earlier that

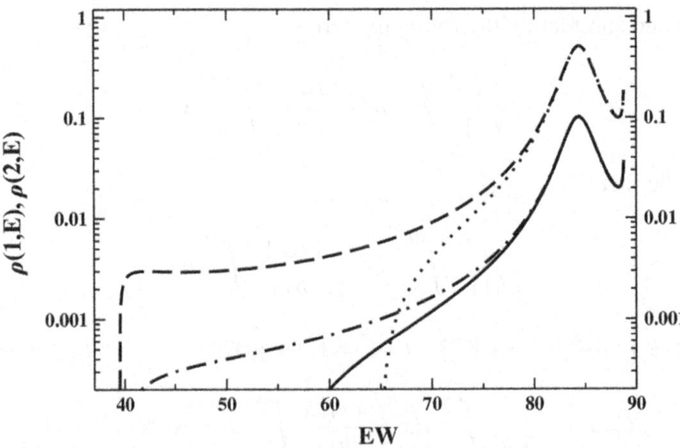

Fig. 4.4 The plot is of injectance versus incident energy EW. We have used $e\gamma W = -15$, and $y_j = 0.45W$. The figure shows that semi-classical formula becomes exact at resonance where there is a peak in injectance. Injectance of first transverse mode is less than that of the second

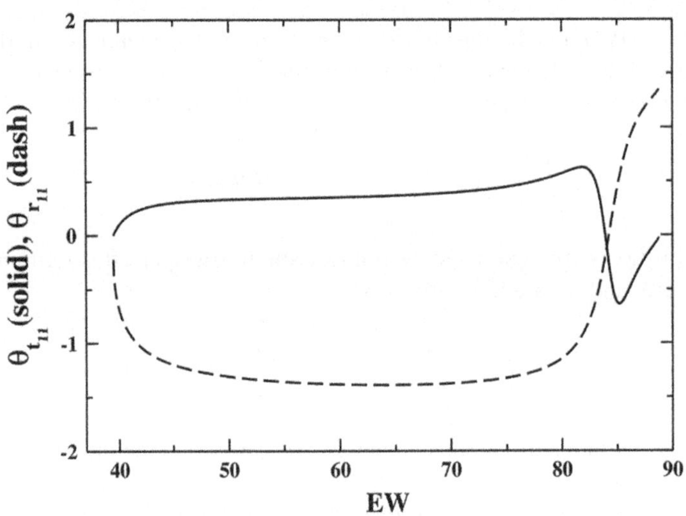

Fig. 4.5 Here, we plot reflection phase shift (dashed curve) and transmission phase shift (solid curve) versus incident energy for the same parameter values as in Fig. 4.4. The figure shows that the scattering phase shifts become simultaneously zero at the resonance while being out of phase for the rest of the energy regime. And so the initial phase of the wavefunction which is usually undetermined do not play a role. That initial phase can be absorbed in $\theta_{r_{11}}$ to give the same conclusion

the exactness of a semi-classical formula can find applications which will be further elaborated in the next two chapters.

It is to be noted that that the semi-classical formula becomes exact (not a good approximation but exact) in a quantum regime which will be proved in the next two sections and physical reasons will be given. But here, we demonstrate in Fig. 4.5, plotted for the same parameters as in Fig. 4.4, the fact that at the resonance $EW = 84.29$ the phase fluctuations maximize. However, at the resonance, the correction term goes to zero as $\theta_{r_{11}}$ goes to zero resulting in $\sin(\theta_{r_{11}})$ becoming zero in Eq. 4.5. The same happens for $\theta_{r_{22}}$. Hence, this situation is different from what was thought earlier that only in the semi-classical regime when $|r_{11}|^2$ becomes negligible, the correction term can be ignored. In a semi-classical regime, particles are expected to behave like a classical particle which is entirely transmitted.

References

1. P.F. Bagwell, Phys. Rev. B **41** 10354 (1990)
2. B.F. Bayman, C.J. Mehoke, Am. J. Phys. **51**(10), 875 (1983)
3. O. Kalman, P. Foldi, M.G. Benedict, F.M. Peeters, Physica E **40**, 567 (2008)
4. W. Porod, Z. Shao, C.S. Lent, Phys. Rev. B **48** 8495 (R) (1993)
5. U. Satpathy, P. Singha Deo, IJMPB **26** 1250028 (2012)
6. P. Singha Deo et al., Phys. Rev. B **58**, 10784 (1998)
7. P. Singha Deo, Phys. Rev. B **75**, 1235330 (2007)

Chapter 5
Negative Partial Density of States

Fano resonance in quantum wires as demonstrated in the previous Chap. 4 is of course well known in optics (Nye and Berry 1974; Berry 1998), but the mesoscopic scenario is different and novel. This difference essentially originates from the distinct sub-band quantization (due to the confinement in y-direction as demonstrated in Fig. 2.8). Such a sub-band quantization has not yet been observed with optical waveguides essentially because of the much shorter wavelength of light. Besides phase changes are very important in mesoscopic physics as they are related to breakdown of parity effect, interpretation of Friedel sum rule (FSR), relation to partial density of states (PDOS), etc., that determine the thermodynamic properties of a mesoscopic system which is absent in case of optics. In this chapter, we discuss the scattering phase shifts and their discontinuities from the point of view that is employed in optics and we will of course be motivated by parity effect and FSR-like laws. This approach is related to the argument theorem in complex analysis and referred to as Burgers circuit (BC). We will show that the BC has a deep connection to density of states and partial density of states which was unknown before our works. This will help us prove the reality of negative PDOS and prove the general regimes when semi-classical FSR can be exact in a purely quantum regime. The advantage of using BC is that these results can be shown to be very general and thus goes beyond all earlier works (Satpathy and Singha Deo 2016).

Besides, we will show that the scattering phase shift behavior of different experimentally and theoretically studied systems can be understood from the Argand diagrams (A-Ds) analyticity. Each such system has its own peculiarities and so it is important to understand them with respect to a mathematical principle like BC. A-D for the scattering matrix element of these systems can be classified as (a) A-D is closed, (b) A-D is open, (c) A-D encloses the phase singularity, (d) Argand diagram does not enclose the phase singularity, (e) A-D is simply connected and (f) A-D is multiply connected due to the presence of subloops. Specific properties of the scatterer only matter to the extent that the A-D changes from one of these

P. Singha Deo, *Mesoscopic Route to Time Travel*,
https://doi.org/10.1007/978-981-16-4465-8_5

to another. Among them, some changes are topologically possible and others are not. Understanding how an A-D changes from one to the other explains most of the puzzles.

In Sect. 5.1, we will give an elementary introduction to BC. In Sect. 5.2, we will analyze slips in the scattering phase shift for a few potentials. In Sect. 5.3, we will show that FSR too can be understood from A-Ds although its manifestation seems to be completely different for different potentials. We will also show why FSR can become exact whenever there is a phase lapse. This is a physically counter-intuitive result that has been proven for particular potentials so far. We will show that this result depends on the properties of A-D and hence very general and independent of the scattering potential. In Sect. 5.3, we will also use Burgers circuit to prove the possibility of negative partial density of states in real mesoscopic systems. To show negative PDOS by explicit calculation for any particular realistic potential is virtually impossible and so never shown before. In Sect. 5.4, we present a brief summary of this chapter.

5.1 Burgers Circuit: An Introduction

A-D is a plot of real versus imaginary parts of an analytic complex function and BC is about phase changes and lapses being determined by phase singularities. Given a complex function t if there is a phase singularity in its complex plane, I is a topological quantum number which is always conserved in an interaction. I is a sign and so can be $+1$, -1 and 0. For a generalized 'Burgers circuit' (BC) (Berry 1998),

$$\oint_C d\phi = 2\pi I \tag{5.1}$$

where $\phi = Arctan\frac{Im(t)}{Re(t)}$. If the contour C does not enclose the phase singularity then I is 0. When the contour C enclosing a phase singularity is clockwise then I is -1 and when the contour C enclosing a phase singularity is anticlockwise then I is $+1$. A scattering matrix element is a complex function for which A-D can be drawn and concept of BC can be applied. For the rest of this chapter where we refer to scattering matrix element, we will mean the scattering matrix element of a quantum mechanical particle say an electron. A-D for a scattering matrix element is generally anticlockwise, as incident energy of the scattering particle increases. At the phase singularity, $Re(t) = 0$ and $Im(t) = 0$, implying the phase singularity is at the origin. Figure 5.1a, b, shows schematic A-Ds for a complex function t. Figure 5.1a, shows a typical anticlockwise contour (ABDEA) of an A-D enclosing the phase singularity at the origin. Hence, for this case, $I = +1$. The contour trajectory (ABDEA) is concave throughout with respect to the singular point at the origin. Following Eq. (5.1), the net change in phase in tracing ABDEA, in Fig. 5.1a, is 2π. Thus, the phase monotonously increases for a concave trajectory. Figure 5.1b shows a typical contour (FGHJF) of

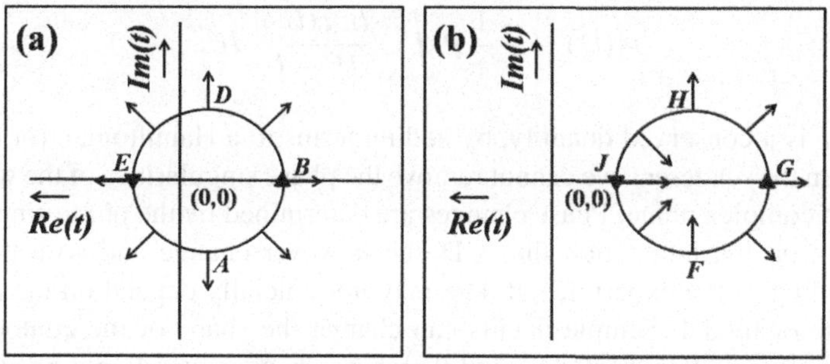

Fig. 5.1 Schematic A-Ds that exemplify Eq. 5.1. **a** When the A-D contour encloses the singular point at (0, 0) and **b** when the contour does not enclose the singular point

an A-D not enclosing the phase singularity at the origin. Hence, for this case, $I = 0$. The contour has both concave (FGH) and convex (HJF) trajectories with respect to the singular point at the origin. Following Eq. (5.1), the total change in phase in Fig. 5.1b is zero. It is possible if the phase increases for the concave trajectory and decreases for the convex trajectory, the net phase change being zero. This can be easily verified by calculating ϕ at any point $(Re(t), Im(t))$ of the trajectory and will be further demonstrated below. Thus, for both Fig. 5.1a, b, the phase change is determined by the singular point and the two follow the same principle. As a special case of the two, one can have a situation where the contour touches the singular point. In which case too, the phase change can be understood from the same principle and will be explained later. Also if it is a closed contour as in Figs 5.1a, b, then Eq. (5.1) is exact. In the real systems, we will discuss in this work, the contours may not be closed. However, we will extend Eq. (5.1) to understand such cases too.

Whatever be the shape of a closed contour C, the phase change is given by Eq. (5.1). As a consequence, the real and imaginary parts of t are not independent of each other but are related. Let us say the complex transmission amplitude be $t(U) = |t(U)|e^{i\theta_t(U)}$, where U can be any parameter like incident energy or gate voltage. Then

$$Re\,(t(U)) = \frac{1}{\pi}P\int_{-\infty}^{\infty}\frac{Im(t(U'))}{U'-U}dU' \tag{5.2}$$

$$Im\,(t(U)) = -\frac{1}{\pi}P\int_{-\infty}^{\infty}\frac{Re(t(U'))}{U'-U}dU' \tag{5.3}$$

These are the well-known Kramers-Kronig relations. Another way in which the relation can be stated is in terms of Hilbert transform,

$$ln|t(U)| = \frac{1}{\pi}P\int_{-\infty}^{\infty}\frac{\theta_t(U')}{U'-U}dU' \tag{5.4}$$

$$\theta_t(U) = -\frac{1}{\pi} P \int_{-\infty}^{\infty} \frac{ln|t(U')|}{U' - U} dU' \quad (5.5)$$

Since I is a conserved quantity, by adding terms to a Hamiltonian (or details to the states in the scatterer), we cannot remove the phase singularities of the wavefunction in the complex plane. Phase changes are determined by the phase singularities. Depending on the interaction the A-D can however change and so a theoretical understanding of the experimental data may not crucially depend on the details of the sample or model. Sample details can change the shape of the contour C, but as these theorems state, to understand the phase changes, we do not need all these details. Englman and Yahalom (2000) had shown that the experimental data for scattering phase shift and scattering cross section of a quantum dot, are consistent with Hilbert transforms. We will show that the principles of analyticity and Eq. (5.1) can be used to establish our results rigorously (Satpathy and Singha Deo 2016).

5.2 Model Potentials

In this section, we analyze scattering phase shifts for the different potentials with respect to Eq. (5.1). As will be clear, we do not need very complicated realistic potentials to understand the phase shifts but we need representative potentials that can be exactly solved and help us understand different aspects of Eq. (5.1).

5.2.1 Double Delta Function Potential in One Dimension

Let us first consider scattering by a double delta function potential in one dimension (1D) schematically shown in Fig. 5.2a. Although a simple potential, it exhibits pronounced Breit-Wigner (BW) resonances. We will use Eq. (5.1) to understand the scattering phase shift for this system and hence for BW resonances. The scattering potential for this system can be written as

$$V_1(x) = \gamma_1 \delta(x)$$

$$V_2(x) = \gamma_2 \delta(x - a)$$

where γ_1 and γ_2 are the strengths of potential V_1 and V_2, respectively. The wavefunction in different regions marked I, II and III are (see Fig. 5.2a),

$$\psi(x) = \begin{cases} e^{ikx} + re^{-ikx}, for\, x < 0, \\ Ae^{ikx} + Be^{-ikx}, for\, 0 \leq x \leq a, \\ te^{ik(x-a)}, for\, x > a. \end{cases}$$

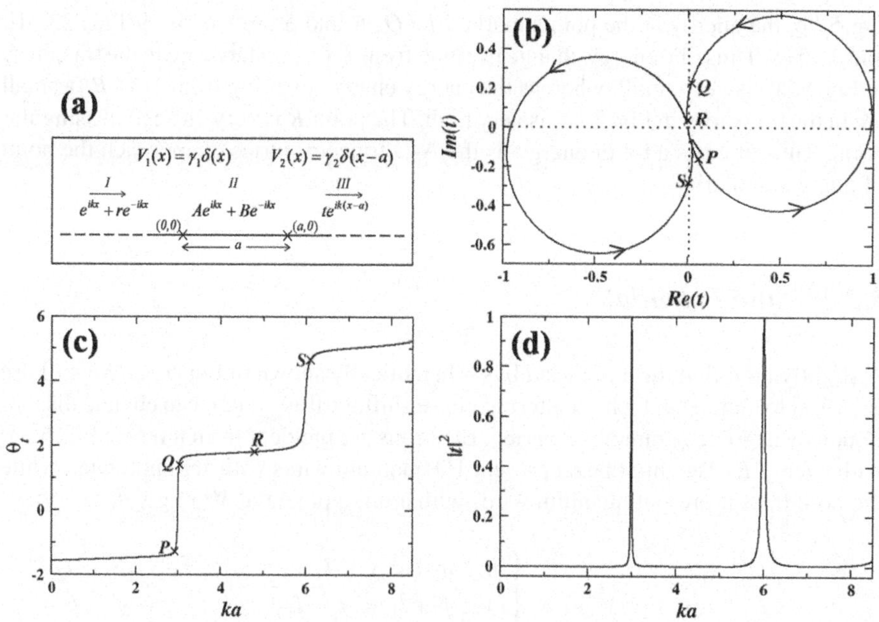

Fig. 5.2 **a** Schematic representation for scattering of electrons by a double delta function potential in 1D. The direction of incident and transmitted electrons are represented by arrows. The solid line represents a quantum wire with double delta function potentials at positions $x = 0$ and $x = a$, respectively, shown by cross (X) marks. γ_1 and γ_2 are the strengths of the potentials. The dashed lines represent the fact that the quantum wire is connected to electron reservoirs via leads. **b** A-D for transmission amplitude for the double delta function potential. **c** Plot of transmission phase shift θ_t versus ka and **d** plot of transmission coefficient $|t|^2$ versus ka, for the double delta function potential using parameters $e\gamma_1 a = e\gamma_2 a = 40$, $a = 1$, $e = 1$, $2m_e = 1$ and $\hbar = 1$

Here, r and t are the reflection and transmission amplitudes, respectively, $k = \sqrt{\frac{2m_e}{\hbar^2}E}$ is the wavevector and E is incident Fermi energy. $t =| t | e^{i\theta_t}$, where $\theta_t = Arctan\frac{Im(t)}{Re(t)}$ is transmission phase shift and $|t| = \sqrt{Im(t)^2 + Re(t)^2}$ is transmission modulus. The A-D for t is shown in Fig. 5.2b, where energy is varied to remain within the first Riemann surface. There is a phase singularity at the origin where $t = 0$. The A-D encloses the singularity but is not closed in the first Riemann surface. Figure 5.2c, d shows the transmission phase shift θ_t and the transmission coefficient $|t|^2$, respectively, as a function of ka, using the same parameters as in Fig. 5.2b.

In Fig. 5.2b, the contour starts from the origin where $E = 0$, goes first through point P and then through Q, R and S. The trajectory facing the singular point at the origin is concave throughout, and thus as discussed with Fig. 5.1, the phase increases continuously. This is evident in Fig. 5.2c where the points P, Q, R and S are also shown at their respective values of ka. As the trajectory comes closer to the point of phase singularity, phase changes are very small and energy cost is very high. In

Fig. 5.2b, the energy at the points marked P, Q, R and S are $8.614a$, $9.12a$, $23.61a$ and $37.58a$. Thus, the energy change in going from P to Q (a large arc in the trajectory in Fig. 5.2b) is very small, whereas the energy change in going from Q to R (a small arc in the trajectory in Fig. 5.2b) is very high. The point R is very close to the singular point. Thus, it costs a lot of energy as the A-D trajectory tries to approach the point of phase singularity.

5.2.2 Stub Potential

A slightly modified form of the stub is schematically shown in Fig. 5.3a. We will use Eq. (5.1) to understand why scattering phase shift for this system can change discontinuously as some parameter is varied. Electrons are incident from left (see Fig. 5.3a) with energy E. The thin lines represent 1D quantum wires with zero potential, while the bold lines represent quantum wires with a finite potential $V(y)$ given by

$$V(y) = \begin{cases} 0, \ for\, 0 < y \leq l_1, \\ iV_0, \ for\, l_1 < y < l, \\ \infty, \ for\, y \geq l. \end{cases}$$

The potential $V(y)$ is taken to be imaginary, as it allows us to make the A-D trajectory approach and cross the point of phase singularity. This cannot be done with real potentials as I in Eq. (5.1) is a conserved quantity. Imaginary potentials are known as optical potentials (Jayannavar 1994). They are often used to simulate the effect of decoherence and open systems. If the system changes from a closed one to an open one then I may be different for the two cases. When V_0 is 0 then $V(y)$ is 0 and the system is a closed conserved system. And when V_0 is non-zero then $V(y) = iV_0$ and it is an open system. The wavefunction in the different regions are given by

$$\psi(x, y) = \begin{cases} e^{ikx} + re^{-ikx}, \ for\, x < 0, \\ te^{ikx}, \ for\, x > 0, \\ Ae^{iky} + Be^{-iky}, \ for\, 0 < y \leq l_1, \\ Ce^{iq(y-l_1)} + De^{-iq(y-l_1)}, \ for\, l_1 < y < l, \\ 0, \ for y = l. \end{cases}$$

Here, r and t are the reflection and transmission amplitudes, respectively, $k = \sqrt{\frac{2m_e}{\hbar^2}E}$ is the wavevector along thin lines, $q = \sqrt{\frac{2m_e}{\hbar^2}(E - iV_0)}$ is the wavevector along the bold line and E is the Fermi energy. Solving the scattering problem using boundary conditions elaborated in Chap. 1 for Fig. 5.3a, we get r and t as a function of energy, E.

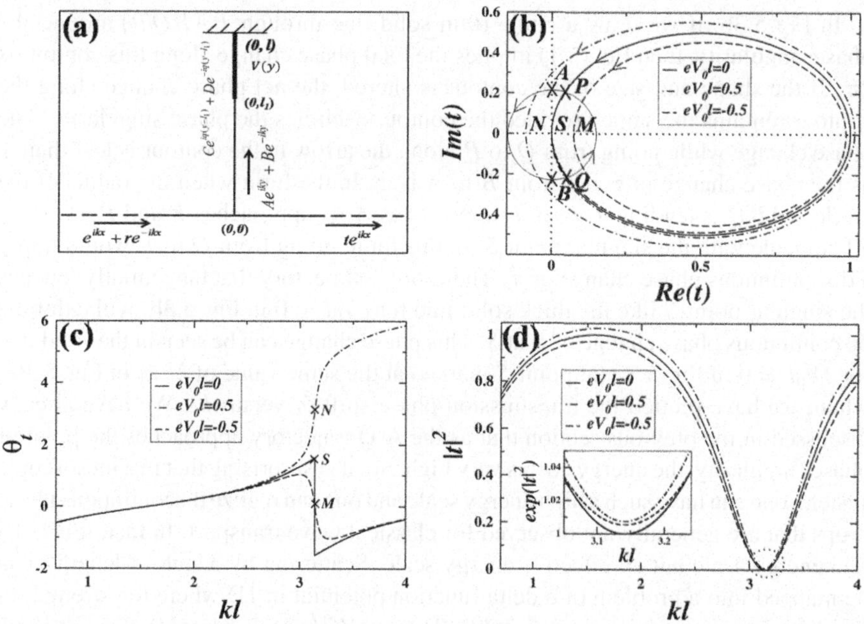

Fig. 5.3 **a** Schematic representation for scattering of electrons by a stub potential. The direction of incident and transmitted electrons are shown by arrows. The potential, represented by the bold line along y-axis, is $V(y) = iV_0$. **b** A-D for transmission amplitude for different values of $eV_0 l$, where e is electronic charge and l is the length shown in (**a**). The thick solid line is A-D for the case when $eV_0 l = 0$, the dot-dashed line is that for $eV_0 l = 0.5$ and the dashed line is for $eV_0 l = -0.5$. Here, $l_1 = .5l$, $l = 1$, $e = 1$, $2m_e = 1$ and $\hbar = 1$. **c** Plot of transmission phase shift θ_t and **d** plot of transmission coefficient $|t|^2$, as a function of dimensionless wavevector kl, taking the same notations and parameters as in Fig. 5.3b

$$r = \frac{1 - \frac{ik-q_1}{ik+q_1} e^{-2ikl_1}}{1 + 3\frac{ik-q_1}{ik+q_1} e^{-2ikl_1}} \tag{5.6}$$

$$t = \frac{2(1 + \frac{ik-q_1}{ik+q_1} e^{-2ikl_1})}{1 + 3\frac{ik-q_1}{ik+q_1} e^{-2ikl_1}} \tag{5.7}$$

where $q_1 = q\,[cot\,(q(l - l_1))]$. Transmission phase shift is given by $\theta_t = Arctan\frac{Im(t)}{Re(t)}$.

Figure 5.3b show the A-Ds for the transmission amplitudes, given by Eq. (5.7), for different values of $eV_0 l$, where energy is varied to remain within the first Riemann surface. The thick solid line is for $eV_0 l = 0$, the dot-dashed line is for $eV_0 l = 0.5$ and the dashed line is for $eV_0 l = -0.5$. There is a phase singularity at $t = 0$ marked as S in Fig. 5.3b. Figure 5.3c, d show the transmission phase shift θ_t and transmission coefficient $|t|^2$, respectively, as a function of kl for different values of $eV_0 l$, using the same parameters and same notations as in Fig. 5.3b.

In Fig. 5.3b, if we draw a circle (thin solid line through $PABQP$) around the phase singularity, then Eq. (5.1) implies the total phase change along this contour is 2π. If the shape and size of the contour is altered, the net phase change along the contour remains the same provided the contour encircles the phase singularity. The phase change while going from Q to P along the arrow in the contour is less than π as the phase change in going from B to A is π. In the limit when the radius of the circle $PABQ$ is tending to zero, P approaches A, Q approaches B and P, A, B, Q all coincide with the singular point S. In this limit, going from Q to P would imply a discontinuous phase change of π. Therefore, a trajectory that tangentially touches the singular point S like the thick solid line for $eV_0l = 0$ in Fig. 5.3b, will exhibit a discontinuous phase change of π at S. This phase change can be seen in the solid line for $eV_0l = 0$ in Fig. 5.3c (at point S marked at the same value of kl as in Fig. 5.3b), where we have plotted the transmission phase shift θ_t versus kl. We have already discussed in the previous section that as the A-D trajectory approaches the point of phase singularity, the energy cost is very high. So, it is surprising that in a mesoscopic system, one can have such a high energy scale and one can realize discontinuous phase drops that are generally not observed for classical wave transport. In fact, it is not a real energy scale but an effective energy scale. Scattering by a stub of length l can be mapped into a problem of a delta function potential in 1D where the strength of the delta function potential is $kcot(kl)$. That is, $V^{eff}(x) = kcot(kl)\delta(x)$ has same reflection amplitude r and transmission amplitude t as a stub of length l. This is an effective potential that the electrons encounter, while the real potential is $V_0 = 0$. At $kl = \pi$, a small change in k or l means a very large change in the effective potential $V^{eff}(x)$. Such large effective energy scales are known in other areas of condensed matter physics, for example, the effective electron mass becomes ∞ at the band edge.

For $eV_0l = 0.5$, the A-D trajectory shown by the dot-dashed line in Fig. 5.3b, intersects the $Re(t)$-axis at point N at $kl = \pi$. The A-D trajectory facing the phase singularity at S is concave throughout. So, Eq. (5.1) implies a monotonously increasing phase. This phase behavior can be seen in the dot-dashed line in Fig. 5.3c where we have plotted the transmission phase shift θ_t versus kl and marked N at $kl = \pi$. For $eV_0l = -0.5$, the A-D trajectory shown by the dashed line in Fig. 5.3b, intersects the $Re(t)$-axis at point M at $kl = \pi$. The A-D trajectory facing the phase singularity is partially convex and partially concave. As discussed with respect to Eq. (5.1), the phase increases for the concave part and decreases for the convex part of the trajectory. This phase behavior can be seen in the dashed curve in Fig. 5.3c where we have plotted the transmission phase shift θ_t versus kl and marked the point M similarly. Therefore, if eV_0l is continuously changed from 0.5 to −0.5, then the point N moves to M crossing the point of phase singularity. When eV_0l becomes 0 we get the solid curves in Fig. 5.3b–d. This implies that with an imaginary potential, just by changing sign of the potential it is possible to cross the singular point. I for dot-dashed line is 1 and that for the dashed line is 0. In other words by changing an imaginary potential, we can make I change from 1 to 0. This is difficult with a real potential. If we tried to cross the singular point with real potentials it would cost an infinite amount of energy. We have thus discussed an effective real potential that

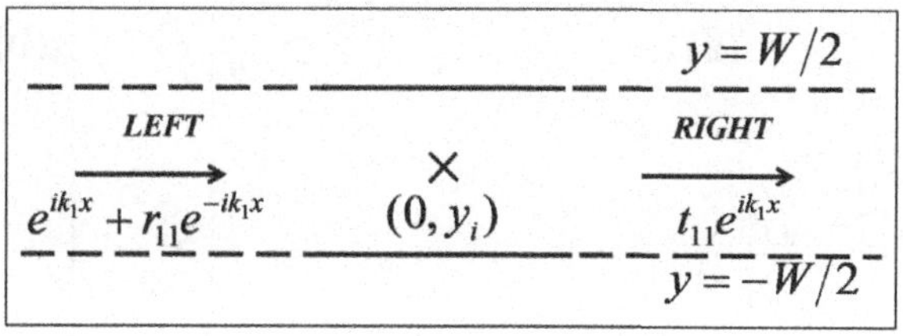

Fig. 5.4 Schematic representation for scattering of electrons by a delta function potential in Q1D. The position of a delta function potential is shown by cross (X) mark

can make an A-D trajectory tangentially touch the point of phase singularity. Any effective potential that can make A-D trajectory cross the point of phase singularity is not known.

In Fig. 5.3c, we observe two different types of phase drops. For $eV_0l = 0$, we get a discontinuous phase drop (shown by the solid curve in Fig. 5.3c) and for $eV_0l = -0.5$, we get a gradual phase drop (shown by the dashed curve in Fig. 5.3c). If we make $eV_0l < -0.5$, the point M in Fig. 5.3b will shift more to the right and the phase drop will be less in magnitude and also less sharp. In Fig. 5.3d, using the same parameters and same notations as in Fig. 5.3b, c, transmission coefficient $|t|^2$ is plotted as a function of kl for different values of eV_0l. The inset in Fig. 5.3d shows the transmission coefficient $|t|^2$ in exponential scale in the region around $kl = \pi$ (shown by dotted circle). At $kl = \pi$, the thick solid curve for $eV_0l = 0$ goes to zero, while the dot-dashed and dashed curves for $eV_0l = 0.5$ and $eV_0l = -0.5$, respectively, go through a non-zero minima and are very close. The phase behaviors for dot-dashed and dashed curves are completely different as can be seen from Fig. 5.3c.

Thus, in this section, we have explained using A-D and BC, why phase (scattering phase shift) drops can occur discontinuously? Why such a phase drop can disappear or change from discontinuous to gradual? The gradual phase drop in the dashed curve of Fig. 5.3c at M is related to the analytic property of complex transmission amplitude and how the trajectory encloses the singularity. Whether the gradual drop is sharp or slow, depends on the distance SM (Fig. 5.3b) at which the convex trajectory intercepts the $Re(t)$-axis.

5.2.3 Delta Function Potential in Q1D

The next scattering potential we consider, is a delta function potential in a single-channel quantum wire. This essentially means, there is a single propagating channel, while all other channels are evanescent as elaborated in Chap. 4. These evanescent channels are characteristic of quasi-one dimension (Q1D), and make this scattering potential completely different from that of a delta function potential in 1D. This system gained relevance with respect to some recent experiments (Schuster et al. 1997; Singha Deo 1998). The scattering solution for this potential has been discussed in detail in Sect. 4.2 and so we directly proceed to analyze the scattering phase shift in terms of A-Ds.

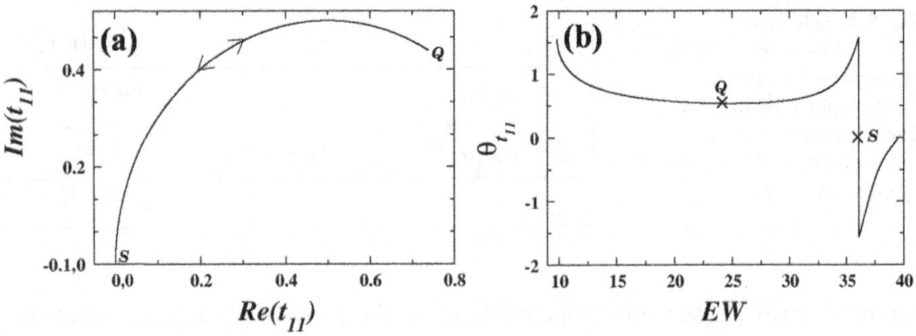

Fig. 5.5 **a** A-D for transmission amplitude and **b** plot of transmission phase shift $\theta_{t_{11}}$ versus EW, for scattering by a delta function potential in a single-channel quantum wire of width W. Here $e\gamma W = -1.5$, $y_i = 0.21W$, $e = 1$, $W = 1$ and we have considered 500 evanescent modes

The A-D for transmission amplitude t_{11} is shown in Fig. 5.5a. There is a phase singularity at the origin where $t_{11} = 0$ (shown by the point S). Figure 5.5b shows transmission phase shift $\theta_{t_{11}}$ as a function of energy, for the same parameters as in Fig. 5.5a. In Fig. 5.5a, the trajectory (shown by thick solid line) starts from the origin (point marked S) goes up to point Q, and traces back the same path to pass the origin making SQS a closed contour. The direction of the trajectory is therefore shown by a double-headed arrow. At the energy, where the trajectory goes from Q to S and touches the point of phase singularity at origin, i.e., point S, we expect a discontinuous phase drop of π, following the same argument as in the case of the stub. This phase drop can be seen in Fig. 5.5b, where the points Q and S are also marked at their respective energies. Thus, the discontinuous phase drop is a natural consequence of Eq. (5.1).

5.2.4 Three Prong Potential

We now consider another potential called the three prong potential. This potential will help us to demonstrate other non-trivial aspects that follow from Eq. (5.1). A schematic representation of the three prong potential is shown in Fig. 5.6. The thin lines represent 1D quantum wires with potential $V = 0$, and the bold lines represent quantum wires with non zero potential, i.e., $V \neq 0$. The arms of the prong are labeled as 1, 2 and 3 as shown in Fig. 5.6. The electrons are considered to be incident from left, the direction of incidence being shown by arrows. The wavefunction in the different regions are given by

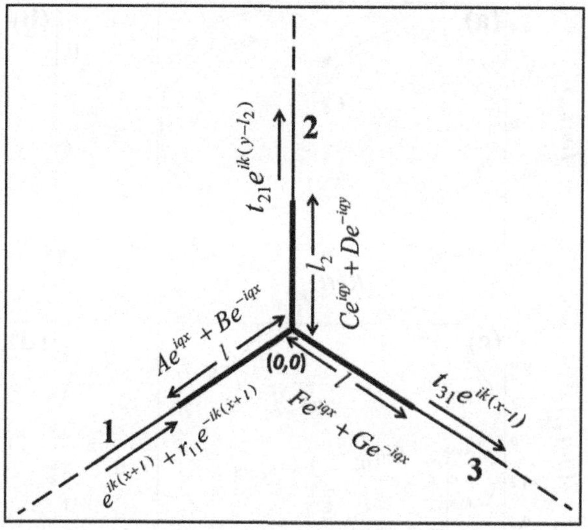

Fig. 5.6 Schematic representation of scattering of electrons by a three prong potential. The direction of incident and transmitted electrons are shown by arrow heads. The potential is non-zero along the bold lines of lengths l, l_2 and l along $-x$, $+y$ and $+x$ axes, respectively

$$\psi(x,y,z)=\begin{cases} e^{ik(x+l)}+r_{11}e^{-ik(x+l)}, \; for\, x<-l,\\ Ae^{iqx}+Be^{-iqx}, \; for\, -l<x<0,\\ Ce^{iqy}+De^{-iqy}, \; for\, 0<y<l_2,\\ Fe^{iqx}+Ge^{-iqx}, \; for\, 0<x<l,\\ t_{21}e^{ik(y-l_2)}, \; for\, y>l_2,\\ t_{31}e^{ik(x-l)}, \; for\, x>l. \end{cases}$$

where $k=\sqrt{\frac{2m_e}{\hbar^2}E}$ is the wavevector along the thin lines, $q=\sqrt{\frac{2m_e}{\hbar^2}(E-V)}$ is the wavevector along the bold lines and E is the Fermi energy. Here r_{11} is the reflection amplitude for electrons incident from 1 and reflected back to 1, t_{21} is the transmission amplitude for electrons incident from 1 and transmitted to 2 and t_{31} is the transmission amplitude for electrons incident from 1 and transmitted to 3. These scattering matrix elements can be solved using boundary conditions elaborated in Chap. 1. The respective transmission phase shifts are given by, $\theta_{r_{11}}=Arctan\frac{Im(r_{11})}{Re(r_{11})}$, $\theta_{t_{21}}=Arctan\frac{Im(t_{21})}{Re(t_{21})}$ and $\theta_{t_{31}}=Arctan\frac{Im(t_{31})}{Re(t_{31})}$.

Figure 5.7a shows the A-D for transmission amplitude t_{21}. There is a phase singularity at the origin, where $t_{21}=0$. In Fig. 5.7a the trajectory of A-D for t_{21} starts from the origin, goes through P and then through Q, R and S following anticlockwise direction shown by arrows. The trajectory is concave throughout and Eq. (5.1) implies monotonously increasing phase. This monotonously increasing phase can be seen in Fig. 5.7b, where scattering phase shift $\theta_{t_{21}}$ is plotted as a function of kl and here also the points P, Q, R and S are marked at the corresponding values of kl.

The A-D for t_{31} shows something interesting. This A-D is shown in Fig. 5.7c. There is a phase singularity at the origin, where $t_{31}=0$. In Fig. 5.7c, the trajectory

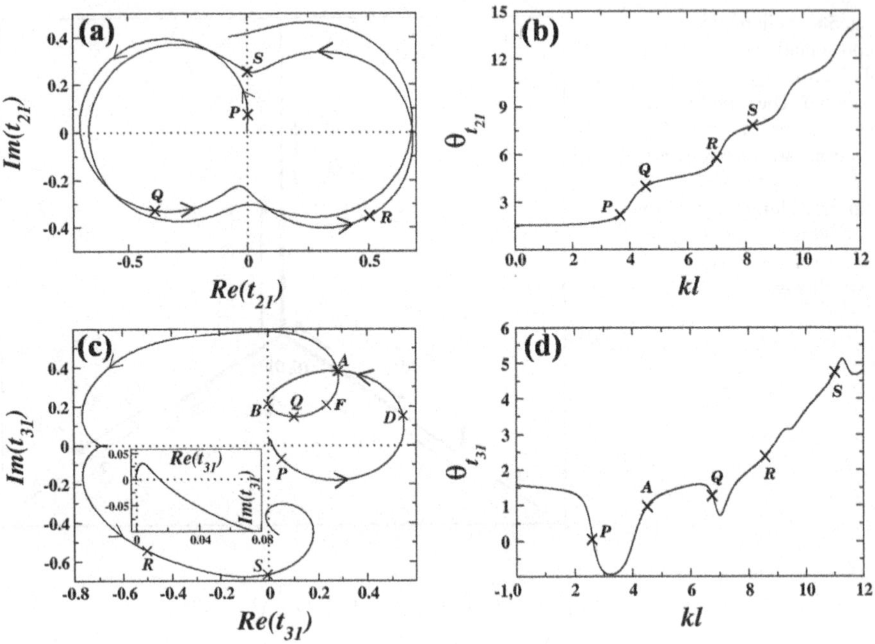

Fig. 5.7 **a** A-D for transmission amplitude t_{21} and **b** plot of scattering phase shift $\theta_{t_{21}}$ as a function of kl. **c** A-D for transmission amplitude t_{31} and **d** plot of scattering phase shift $\theta_{t_{31}}$ as a function of kl. For all the figures, $l = 1$, $l_2 = 5l$, $e = 1$ and $eVl = -1000$

of A-D for t_{31} starts from origin, goes through P and then through D, A, B, Q, F, A and S, following anticlockwise direction shown by arrows. Here interestingly, the trajectory develops a subloop $ABQFA$. This subloop results in a convex arc BQF in the trajectory that does not go through the origin. As explained earlier, there will be a gradual phase drop whenever such a convex arc is observed, following Eq. (5.1). This can be seen in Fig. 5.7d, where scattering phase shift $\theta_{t_{31}}$ is plotted as a function of kl and here also the points P, A, Q, A and S are marked. The presence or absence of such a subloop has no consequence on the line integral of phase along $PDABQFAS$. This is because the contribution to the line integral coming from the subloop $ABQFA$ is 0 and its presence or absence has no bearing on the value of I. So, such a subloop as $ABQFA$ in Fig. 5.7c can appear or disappear (in the sense that it can reduce to a cusp) as some parameter is varied as will be demonstrated in the next section.

5.3 Injectance and Friedel Sum Rule

Note substituting $s_{\alpha\beta} =| s_{\alpha\beta} | e^{i\theta_{s_{\alpha\beta}}}$, PDOS in Eq. 3.10 can be written as

$$\rho_{pd}(E, \alpha, \beta) = -\frac{1}{2\pi} \int_{sample} d\mathbf{r}^3 \left(|s_{\alpha\beta}|^2 \frac{\delta\theta_{s_{\alpha\beta}}}{\delta V(\mathbf{r})} \right) \tag{5.8}$$

Here, the region of integration is the "sample" which was earlier denoted by Ω, $s_{\alpha\beta}$ is the scattering matrix element for electrons incident from channel β and transmitted to channel α and $\frac{\delta}{\delta V(\mathbf{r})}$ stands for a functional derivative with respect to the local potential $V(\mathbf{r})$. PDOS are quite physical and manifests in a variety of experimental situations in mesoscopic systems (Buttiker 2001). For example, decoherence in the scattering region is proportional to the time electrons spend in the scattering region. As another example, consider a sinusoidal voltage of frequency ω, $V_\beta(\omega)$ applied at incident lead β. The current measured at lead α will be,

$$I(\alpha, \omega) = G_{\alpha\beta}(\omega) V_\beta(\omega) \tag{5.9}$$

where, $G_{\alpha\beta}(\omega)$ is the dynamical conductance matrix and is given by,

$$G_{\alpha\beta}(\omega) = G^0_{\alpha\beta} - i\omega E_{\alpha\beta} + K_{\alpha\beta}\omega^2 + O(\omega^3) \tag{5.10}$$

$G^0_{\alpha\beta}$ is the dc-conductance matrix. $E_{\alpha\beta}$ is proportional to ω and governs the displacement currents and is given by,

$$E_{\alpha\beta} = e^2 \rho_{pd}(\alpha, \beta, E) - e^2 \int d\mathbf{r}' \rho_e(\alpha, \mathbf{r}', E) \int d\mathbf{r} g(\mathbf{r}, \mathbf{r}') \rho_i(\mathbf{r}, \beta, E) \tag{5.11}$$

where, $g(\mathbf{r}, \mathbf{r}')$ is the effective interaction potential. All these experimental situations explicitly involve β as the incoming channel and α as the outgoing channel. All analysis of $\rho_{pd}(E, \alpha, \beta)$ will be made w.r.t the R.H.S of Eq. (5.8).

The integration over $\mathbf{r}$ in Eq. 5.8 can easily be done for a global change (for all $\mathbf{r}$ in the sample as well as in the leads) in $V(\mathbf{r})$ by a constant amount ϵ, i.e., $\delta V(\mathbf{r}) = \epsilon$ for all $\mathbf{r}$. Such a constant global increase in potential is equivalent to decrease in incident energy E, i.e.,

$$\int_{global} d\mathbf{r}^3 \frac{\delta}{\delta V(\mathbf{r})} \equiv -\frac{d}{dE} \tag{5.12}$$

and, therefore,

$$\int_{sample} d\mathbf{r}^3 \frac{\delta}{\delta V(\mathbf{r})} \cong -\frac{d}{dE} \tag{5.13}$$

is expected to work in the semi-classical limit. So, from Eq. 5.8 or 3.10

$$\rho_{pd}(E,\alpha,\beta) = -\frac{1}{4\pi i}\int_{sample} d\mathbf{r}^3 \left(s^*_{\alpha\beta}\frac{\delta s_{\alpha\beta}}{\delta V(\mathbf{r})} - s_{\alpha\beta}\frac{\delta s^*_{\alpha\beta}}{\delta V(\mathbf{r})}\right)$$

$$\approx -\frac{1}{4\pi i}\int_{global} d\mathbf{r}^3 \left(s^*_{\alpha\beta}\frac{\delta s_{\alpha\beta}}{\delta V(\mathbf{r})} - s_{\alpha\beta}\frac{\delta s^*_{\alpha\beta}}{\delta V(\mathbf{r})}\right)$$

$$\text{or,}\quad \rho_{pd}(E,\alpha,\beta) \approx \frac{1}{4\pi i}\left(s^*_{\alpha\beta}\frac{d s_{\alpha\beta}}{dE} - s_{\alpha\beta}\frac{d s^*_{\alpha\beta}}{dE}\right) \qquad \text{(Using Eq. (5.12))}$$

On substituting $s_{\alpha\beta} = \mid s_{\alpha\beta} \mid e^{i\theta_{s_{\alpha\beta}}}$ we get,

$$\rho_{pd}(E,\alpha,\beta) \approx \frac{1}{2\pi}\left(|s_{\alpha\beta}|^2 \frac{d\theta_{s_{\alpha\beta}}}{dE}\right) \tag{5.14}$$

We have already shown that $\frac{d\theta_{s_{\alpha\beta}}}{dE}$ can be negative. However, concluding $\rho_{pd}(E,\alpha,\beta)$ to be negative when R.H.S of Eq. (5.14) is negative is completely wrong. Equation (5.14) is an approximate equality depending on whether the system is in a semi-classical regime, and so if R.H.S is negative, the L.H.S is not necessarily negative and may well be strongly positive. It is also generally not possible to numerically verify the negativity of the RHS of Eq. 5.8 because in 1D as we will show it never becomes negative and in Q1D or more it is technically and conceptually (Landauer and Martin 1994) very difficult. At a given point r inside the scatterer one cannot tag an electron to talk about an electron going from β to α and so the integrand is complicated. However, at an abstract level it is well defined as we have justified in Chap. 3, and once we integrate over r it corresponds to observables as we can see in Eq. 5.11. We will show that some general conclusions can be drawn about the R.H.S of Eq. 5.8 using the properties of Argand diagram and BC which is true for any potential in 1D or in Q1D.

Even though PDOS as defined in Eq. 5.8 can be negative, they add up to give the correct DOS which is positive. Using Eq. (5.13) we get the semi-classical limit of Eq. 3.13,

$$\rho(\beta,E) \approx \frac{1}{4\pi i}\sum_{\alpha}\left(s^*_{\alpha\beta}\frac{d s_{\alpha\beta}}{dE} - s_{\alpha\beta}\frac{d s^*_{\alpha\beta}}{dE}\right) \tag{5.15}$$

Summing $\rho(\beta,E)$ over the β independent channels, we can obtain density of states (DOS) $\rho_d(E)$, i.e., from Eq. 3.15,

$$\rho_d(E) = -\frac{1}{4\pi i}\sum_{\alpha\beta}\int_{sample} d\mathbf{r}^3 \left(s^*_{\alpha\beta}\frac{\delta s_{\alpha\beta}}{\delta V(\mathbf{r})} - s_{\alpha\beta}\frac{\delta s^*_{\alpha\beta}}{\delta V(\mathbf{r})}\right) \tag{5.16}$$

and in the semi-classical limit given by Eq. (5.13), we get,

$$\rho_d(E) \approx \frac{1}{4\pi i}\sum_{\alpha\beta}\left(s^*_{\alpha\beta}\frac{ds_{\alpha\beta}}{dE} - s_{\alpha\beta}\frac{ds^*_{\alpha\beta}}{dE}\right) \tag{5.17}$$

Further simplification of R.H.S in Eq. (5.17) gives

$$\pi\rho_d(E) \approx \frac{d}{dE}\theta_f(E) \tag{5.18}$$

This is Friedel sum rule (FSR), where $\theta_f(E) = \frac{1}{2i}log(det[S])$ is the Friedel phase, S is the scattering matrix and $\rho_d(E) = \frac{dN(E)}{dE}$ is density of states. Since injectance of all leads are independent (as shown in Sect. 1.1) while they add up to give DOS, it is important to understand injectance in order to understand FSR. So we will restrict our study to injectance.

As we have shown in Chap. 4 semi-classical regime being expressed by Eq. (5.13) does not seem to be sufficient. Sometimes, FSR (or injectance) is exact at all energies (for example the stub) and sometimes it is exact at an energy where quantum fluctuations dominate. There is a huge amount of system to system variation. We have also proved very generally in Chap. 4 that when the phase drops of π are discontinuous like that in the solid curve in Fig. 5.3c or Fig. 5.5b, then at the energy corresponding to this drop, semi-classical injectance will become exact. But when the phase drops are gradual like the dashed curve in Fig. 5.3c, this has not been proved in general but shown for the particular case of a delta function potential in Q1D. BC will help us to derive a general proof that gradual phase drops imply exactness of semi-classical injectance for any arbitrary potential that can exhibit such a gradual phase drop. We will first prove that in realistic mesoscopic systems one can definitely observe negative PDOS. We will show that the R.H.S of Eq. (5.8) can become negative for a realistic mesoscopic system. We will also show that such a conclusion cannot be drawn in 1D that has been extensively studied before (Landauer and Martin 1994). Also we will show that when there are such negative slopes in scattering phase shift of mesoscopic systems then semi-classical FSR can become exact in a quantum regime. Once again our proofs will depend on A-D and Eq. (5.1) and so our proof will be general and not depend on the specific properties of the scattering potential.

For this we first make the connection between Eq. (5.1) and injectance. When S matrix is 2×2 (for example, the cases of double delta function potential (Fig. 5.2a), stub (Fig. 5.3a), single channel quantum wire (Fig. 5.4)), Eq. (5.18) simplifies to,

$$\frac{d}{dE}\theta_t(E) \approx \pi\rho_d(E) \tag{5.19}$$

Here t is the transmission amplitude and $\theta_t = Arctan\frac{Im(t)}{Re(t)}$. It has been proved in Chap. 4 that this equation is valid even if θ_t is discontinuous with E because θ_t is analytic wherein the R.H.S derivative of θ_t is the same as the L.H.S derivative at the discontinuity.

Suppose when the energy is varied from 0 to E_1, then the A-D for a typical scattering matrix element t traces a closed contour C. Below we will show that for such a case one can state that, Eq. (5.1) takes the form

$$\oint_C d\theta_t = \pi N(E_1) \tag{5.20}$$

where, $N(E_1)$ is number of states (obtained by integrating DOS $\rho_d(E)$ from 0 to E_1) below energy E_1 and is to be identified with the conserved quantity I (see Fig. 2b for the factor 2). We start with intuitive arguments before proving this in general. Comparing with Eq. (5.1) we see $\phi \equiv \theta_t$ and $2I \equiv N(E_1)$. If C happens to be a completely closed contour, Eq. (5.20) is exact as it is equivalent to Eq. (5.1). Any complex function or its phase has to satisfy Eq. (5.1) and a scattering matrix element is no exception provided it is analytic. This analyticity is the basic requirement for Eq. (5.8) (PDOS) or Eq. 3.13 (injectance). However, in most cases of scattering problems, C is not completely closed and when C is not completely closed one cannot expect any conserved quantity. In Fig. 5.2b, there is a phase singularity at the origin and the contour enclosing the singularity is not closed in the first Riemann surface. One can restrict the discussion to the first Riemann surface to understand the injectance. When the contour continue to the second Riemann surface, then the contour integral starts including contribution from the second phase singularity in the second Riemann surface. And then one has to extend the discussions here to include the effect of the second singularity. This does not change the arguments given here except that sometimes the error from the first Riemann surface can cancel the error from the second Riemann surface which need not be a systematic behavior. The contour C in Fig. 5.2b starts from origin with zero energy and ends in the first Riemann surface at point marked as S, where the energy is E_1 (say). It is now known that (see Eq. 5.6 in (Yeyati and Buttiker 2000)),

$$\int_{C'} d\theta_t = \int_0^{E_1} \left[\pi \frac{dN(E)}{dE} - ImTr\left(G^a \frac{\partial F^a}{\partial E} \right) \right] dE \tag{5.21}$$

We will replace C by C', when the contour is not completely closed. G^a is the advanced Greens function and F^a is self energy due to coupling the system with the leads. One can then state that $-\int_0^{E_1} ImTr\left(G^a \frac{\partial F^a}{\partial E}\right) dE$ is the correction term for Eq. (5.20) when contour C' is not closed. This statement can be alternately justified as follows. When the self energy is independent of incident energy, then the contour C is closed as well as the correction term is zero implying Eq. (5.21) becomes Eq. (5.20). For a double delta function potential in 1D the correction term is very important to consider. There, energy dependence of self energy can be seen very easily in the broadening of consecutive resonance peaks (shown in Fig. 5.2d). Therefore, one can refine the statement in Eq. (5.13) to state,

$$\oint_C dE \int_{sample} dr^3 \frac{\delta}{\delta V(r)} = \oint_C dE \left(-\frac{d}{dE} \right) \tag{5.22}$$

where, we generate a closed contour C in the A-D by varying energy E from 0 to E_1. Using Eq. 5.12 on the RHS of Eq. 5.22 we get

$$\oint_C dE \int_{sample} dr^3 \frac{\delta}{\delta V(r)} = \oint_C dE \int_{global} dr^3 \frac{\delta}{\delta V(r)} \tag{5.23}$$

In case of the solid line in Fig. 5.3b, the potential everywhere is real and is 0. It traces a closed contour in the first Riemann surface. Thus, for this system Eq. (5.20) is applicable. Explicit calculations of density of states for the stub show this and so everything is consistent with Eq. (5.1). However, for scattering by a delta function potential (Fig. 5.4) in a single channel quantum wire we get a counter intuitive result. In this case, the contour of the A-D (shown in Fig. 5.5a) is closed in a special way and so we expect Eq. (5.20) to be applicable. But explicit calculations of density of states in Sect. 4.2 show, that is not the case. It has been shown earlier that for this system the fundamental theorem of Büttiker-Thomas-Pretre (BTP) also breaks down (Singha Deo 2005) due to the non-analyticity of scattering matrix elements. This is a consequence of the fact that delta function potential in Q1D incorporates a log divergence in scattering matrix elements. Hence, this is a situation where Eq. (5.1) cannot be applied.

For the three prong potential, shown in Fig. 5.6, the scattering matrix is 3×3 and the correct form of FSR is given by Eq. (5.18). Whenever the scattering matrix has a dimension greater than 2, the connection between FSR in Eq. (5.18) and Eq. (5.1) is not straight forward. However, we can make this connection for each partial density of states (PDOS) and is shown below. As an example, let us consider the A-Ds for the three prong potential shown in Fig. 5.7a, c. None of the A-Ds (e.g., Fig. 5.7a, c) are closed in the first Riemann surface. Let us, for example, consider the Argand diagram for t_{31} tracing a contour C' ($PDABQFARS$ shown in Fig. 5.7c) as energy is varied from 0 to E_1. We can show that (see Appendix A),

$$\int_{C'} d\theta_{t_{31}} \approx 2\pi \int_0^{E_1} \frac{\rho_{pd}(3, 1, E)}{|t_{31}|^2} dE \tag{5.24}$$

Now we can again state that for any closed curve like $ABQFA$ in Fig. 5.7c, Eq. (5.24) is exact. That is,

$$\oint_{ABQFA} d\theta_{t_{31}} = 2\pi \oint_{ABQFA} \frac{\rho_{pd}(3, 1, E)}{|t_{31}|^2} dE \tag{5.25}$$

This statement can also be alternately justified as follows. For a closed contour, L.H.S of Eq. (5.25) is zero. Also for a closed contour R.H.S of Eq. (5.25) will be zero as shown in Appendix B for any $\rho_{pd}(\alpha, \beta, E)$ inside a closed contour integral. Note that in Appendix B we have used the definition of $\rho_{pd}(\alpha, \beta, E)$ as given by R.H.S of Eq. (5.8). Hence, it follows that R.H.S of Eq. (5.25) is also zero, where $\rho_{pd}(3, 1, E)$ or any $\rho_{pd}(\alpha, \beta, E)$ is given by the R.H.S of Eq. (5.8). The arguments below although stated for $\rho_{pd}(3, 1, E)$ is therefore true for any $\rho_{pd}(\alpha, \beta, E)$ inside

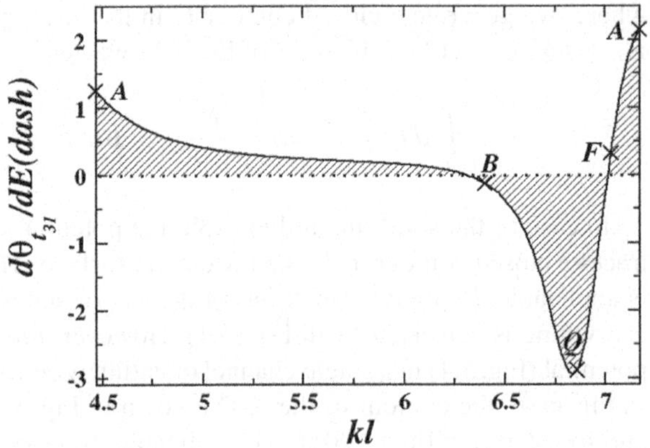

Fig. 5.8 $\frac{d\theta_{t_{31}}}{dE}$ as a function of kl for the three prong potential. Here $l = 1$, $l_2 = 5l$, $e = 1$ and $eVl = -1000$

a closed contour integral. Now,

$$\oint_{ABQFA} d\theta_{t_{31}} = \oint_{ABQFA} \frac{d\theta_{t_{31}}}{dE} dE = 0 \tag{5.26}$$

In Fig. 5.8, $\frac{d\theta_{t_{31}}}{dE}$ is shown in the energy range covering the subloop $ABQFA$ of Fig. 5.7c. As is implied by Eq. (5.26), $\frac{d\theta_{t_{31}}}{dE}$ is somewhere positive and somewhere negative to ensure the area under the curve (shaded region in Fig. 5.8) is zero. Similarly the R.H.S of Eq. (5.25) is zero implies that $\frac{\rho_{pd}(3,1,E)}{|t_{31}|^2}$, will also be positive as well as negative in certain energy values (or kl values). Thus, $\rho_{pd}(3, 1, E)$ as given by the R.H.S of Eq. (5.8) is conclusively negative in some energy values. In some cases the precise energy ranges can be identified and the negativity of the RHS can be further proved as will be shown in the next chapter. Although we have considered the there prong potential as an example the proof is valid for any potential whose A-D shows a continuous phase drop due to a subloop as it is due to the fact that the subloop traces a closed contour.

We will now show how negative slopes in scattering phase shift of mesoscopic systems are fundamentally different from that studied earlier (Landauer and Martin 1994) in 1D. Note the negative slope at point P in Fig. 5.7d. This kind of negative slope at very low energies can arise for scattering in 1D and one can easily check this for a square well potential. In terms of our analysis we understand the negative slope at point P in Fig. 5.7d due to a convex trajectory at P in Fig. 5.7c which is originating due to the fact that the A-D starts from the origin and behaves anomalously as the trajectory starting from the origin is neither clockwise nor anti-clockwise with respect to the singular point (i.e., origin). See the expanded A-D trajectory shown in the inset of Fig. 5.7c. The trajectory moves up, turns around and moves down to become convex in a small energy window and then winds around the origin anti-clockwise. Although $\frac{d\theta_{t_{31}}}{dE}$ is negative at P in Fig. 5.7d, there is no evidence that PDOS $\rho_{pd}(3, 1, E)$ is negative at energy corresponding to point P (to be shown explicitly in next chapter). Such a negative slope is fundamentally different from the negative slope at Q in Fig. 5.7d that originate from a closed subloop $ABQFA$ in Fig. 5.7c, that we encounter only in Q1D and mesoscopic scattering. We have shown that the subloop seen in Fig. 5.7c implies the presence of this negative slope in scattering phase shift $\theta_{t_{31}}$ and also implies PDOS $\rho_{pd}(3, 1, E)$ is negative. We have shown that an A-D for such a scattering matrix element curls around and forms a subloop without violating the topological constraints of Eq. (5.1). The line integration along the subloop $ABQFA$ in Fig. 5.7c does not contribute to the line integration over the trajectory $PDABQFARS$ of Fig. 5.7c or I in Eq. (5.1) is unaffected by the presence or absence of a subloop in the closed contour C. Negative slopes of this second type (i.e., observed at Q in Fig. 5.7d) that we have discussed here in-fact can appear or disappear very easily and can be found at much higher energies. In Fig. 5.9, we have shown the A-D for t_{31} up to very high value of energy (or, kl). We can see many subloops which again implies the presence of negative slopes in scattering phase shift and negative PDOS. Sometimes there is a cusp in the A-D and such a cusp means a subloop has disappeared. Thus, the phase drop will also disappear as we go to such energies. Disappearance of a subloop can be demonstrated by varying any other parameter like V, l, l_2 of the three prong geometry.

Normally one would say that one should start from a closed Hamiltonian system that has well defined quantum states. The system can transmit if the incident energy matches with the energy of the state and transmission characteristics like line shapes and phase shifts are determined by the nature of the states. However, as we can see the system can have a transmission that in no way reflects and is independent of the state inside the system (point P in Fig. 5.7d is a concrete example of it) but rather when it reflects then states are determined by the transmission phase shifts. This gives one the impression that there is an anomalous regime when transmission occurs but states in the system has not developed and so open systems are more general and closed systems with well defined quantum states is a special case. This will be further elaborated in the next chapter.

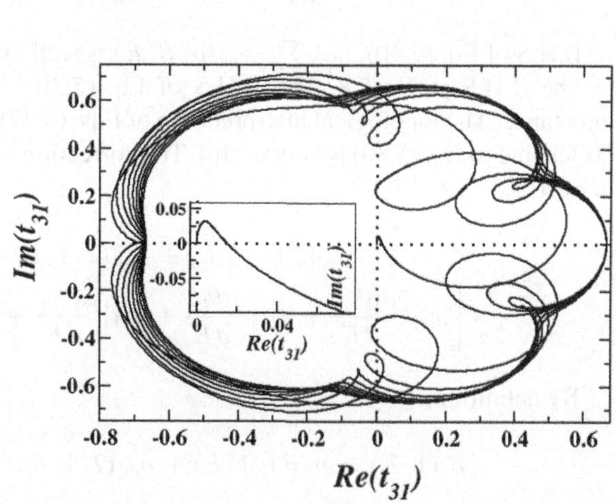

Fig. 5.9 A-D for transmission amplitude t_{31}, varying kl from 0 to 50, for the three prong potential. Here $l = 1, l_2 = 5l, e = 1$ and $eVl = -1000$

$\rho_{pd}(3, 1, E)$ being negative is a counter intuitive feature of quantum mechanics and can have interesting physical significance. Obviously, ac-response of a mesoscopic system will change drastically if $\rho_{pd}(\alpha, \beta, E)$ in Eq. (5.11) changes sign. Also, it means an electron that is incident along lead 1 and transmitted to lead 3, dwells in some negative number of states (PDOS is negative) inside the scatterer. Total charge being electronic charge times number of states will be positive for these negatively charged electrons. So other electrons that are incident along lead 1 and transmitted to lead 2 or reflected back to lead 1 will be attracted by this positively behaving charge of electrons going from 1 to 3. This could be the explanation for the electron-electron attraction observed in the numerical experiment of ref. [P. S. Deo, (2002)], where no explanation could be given.

Having discussed negative PDOS let us come to the puzzling feature of injectance. Let us try to understand if there is a general connection between negative slope and semi-classical injectance becoming exact as observed for specific potentials in Sect. 4.2. The equality in Eq. (5.25) holds for the integrals and does not imply equality of the integrands. However, using Eq. (5.14) we can write

$$\rho_{pd}(3, 1, E) \approx \frac{1}{2\pi}|t_{31}|^2\frac{d\theta_{t_{31}}}{dE} \tag{5.27}$$

$$\rho_{pd}(2, 1, E) \approx \frac{1}{2\pi}|t_{21}|^2\frac{d\theta_{t_{21}}}{dE} \tag{5.28}$$

$$\rho_{pd}(1, 1, E) \approx \frac{1}{2\pi}|r_{11}|^2\frac{d\theta_{r_{11}}}{dE} \tag{5.29}$$

Finally using Eqs. (5.27), (5.28) and (5.29), we can write

$$\begin{aligned}&\rho_{pd}(1, 1, E) + \rho_{pd}(2, 1, E) + \rho_{pd}(3, 1, E) \approx \\ &\frac{1}{2\pi}|r_{11}|^2\frac{d\theta_{r_{11}}}{dE} + \frac{1}{2\pi}|t_{21}|^2\frac{d\theta_{t_{21}}}{dE} + \frac{1}{2\pi}|t_{31}|^2\frac{d\theta_{t_{31}}}{dE}\end{aligned} \tag{5.30}$$

L.H.S of Eq. (5.30), i.e., $\sum_\alpha \rho_{pd}(\alpha, \beta, E)$ is well known as injectance as defined by the R.H.S in Eq. 5.8. The R.H.S of Eq. (5.30) is the semi-classical limit of injectance. The topological interpretation of Eqs. (5.27), (5.28), (5.29) in terms of A-Ds leading to Eq. (5.30) is very useful. The correction term to Eq. (5.30) as discussed in Sect. 3.2

$$\begin{aligned}&\rho_{pd}(1, 1, E) + \rho_{pd}(2, 1, E) + \rho_{pd}(3, 1, E) = \\ &\frac{1}{2\pi}\left[|r_{11}|^2\frac{d\theta_{r_{11}}}{dE} + |t_{21}|^2\frac{d\theta_{t_{21}}}{dE} + |t_{31}|^2\frac{d\theta_{t_{31}}}{dE} + \frac{m_e|r_{11}|}{\hbar k^2}sin(\theta_{r_{11}})\right]\end{aligned} \tag{5.31}$$

By definition,

$$\rho^e(1, E) = \rho_{pd}(1, 1, E) + \rho_{pd}(2, 1, E) + \rho_{pd}(3, 1, E) \tag{5.32}$$

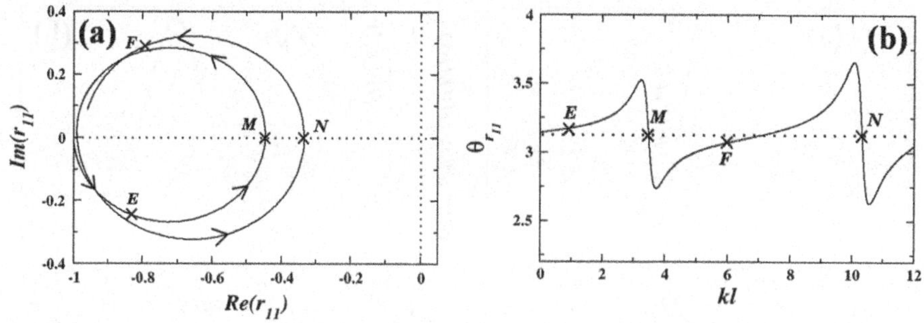

Fig. 5.10 **a** A-D for reflection amplitude r_{11} and **b** plot of reflection phase shift $\theta_{r_{11}}$, versus kl for the three prong potential. Here $l = 1$, $l_2 = 5l$, $e = 1$ and $eVl = -10000$

is exact injectance, defined by the R.H.S in Eq. 5.8. And,

$$\rho^s(1, E) = \frac{1}{2\pi}\left[|r_{11}|^2\frac{d\theta_{r_{11}}}{dE} + |t_{21}|^2\frac{d\theta_{t_{21}}}{dE} + |t_{31}|^2\frac{d\theta_{t_{31}}}{dE}\right] \tag{5.33}$$

is semi-classical injectance. Equation (5.31) implies that $\rho^e(1, E)$ and $\rho^s(1, E)$ will be equal at energies, where the correction term is zero or $|r_{11}|sin(\theta_{r_{11}}) = 0$. According to the earlier arguments of Leavens and Aers described at the end of Chap. 3, in the semi-classical limit $|r_{11}| \longrightarrow 0$ and $\rho^s(1, E) = \rho^e(1, E)$. But in the cases of Fig. 5.10a, $|r_{11}| \neq 0$ at the energy corresponding to the points M and N, where one can see from Fig. 5.10b that $sin(\theta_{r_{11}})$=0 and so $\rho^s(1, E) = \rho^e(1, E)$. We will show how the A-D topology is responsible for this behavior and therefore provides a general understanding. In Fig. 5.10a, b, we have shown the A-D and phase shift for scattering matrix element r_{11}, respectively. kl is varied from 0 to 12 in both the plots. The A-D in Fig. 5.10a is restricted to one side of the phase singularity (i.e., origin) and in the first Riemann surface resulting in subloops. This will naturally mean that, the contour has both concave and convex parts in the trajectory. Scattering phase shift in Fig. 5.10b increases with kl, reaches a peak value and then drops to become π at M. The pattern repeats as kl increases and the scattering phase shift becomes π again at N. Therefore, the correction term $|r_{11}|sin(\theta_{r_{11}})$ to semi-classical injectance (Eq. (5.31)) is zero at M and N. Thus, at M and N, the semi-classical injectance (Eq. (5.33)) is exact. The exactness of semi-classical injectance is shown in Fig. 5.11 (for clarity see the inset) at points M and N corresponding to same kl values as in Fig. 5.10. For monotonously increasing phase, A-D extends to higher Riemann surfaces and line integrals include the effect of other singularities in higher Riemann surface. Of course phase can be integral multiples of π (i.e., 2π, 3π, .. where ρ^s can become exact because the error from one Riemann surface cancels the error from the next Riemann surface), but for open A-D trajectories as argued before, this is not a consistent behavior, except in some simple 1D scattering problems. This inconsistent behavior can be for example, checked for an 1D Aharonov-Bohm ring with different arm lengths. Drops in $\theta_{r_{11}}$ resulting in $\theta_{r_{11}}$ being π and $sin(\theta_{r_{11}}) = 0$, leading to

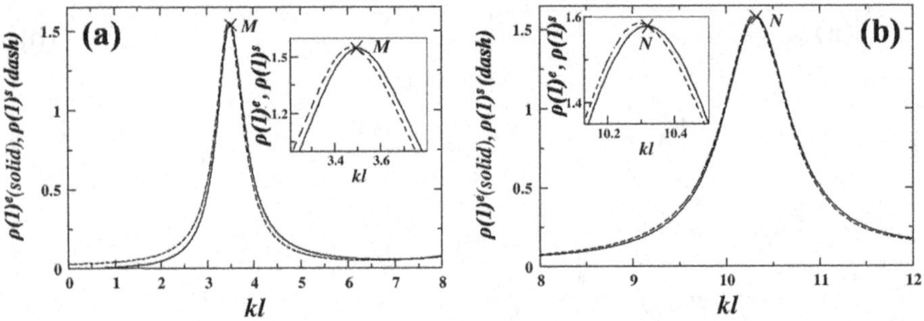

Fig. 5.11 Plot of exact injectance $\rho^e(1, E)$ (solid line) and semi-classical injectance $\rho^s(1, E)$ (dashed line) as a function of kl for the three prong potential. The peaks in the injectance are shown separately, **a** shows the first peak, **b** shows the second peak, for the same parameters as in Fig. 5.10. The insets show the magnified curves at points M and N

semi-classical injectance being exact can be understood from the A-D in a single Riemann surface, involving a single phase singularity in the line integral and hence is not an accident. Drops in $\theta_{r_{11}}$ is a pure quantum mechanical behavior and hence exactness of semi-classical injectance at the energies corresponding to the phase drops, is counter-intuitive. As argued before, these drops coming from subloops in A-D are tunable and can be removed by varying some parameter. In Fig. 5.12, we plot the phase behavior for the same system, after decreasing the potential V of the same system as in Figs. 5.10 and 5.11. The phase drop at point F as usual decreases to π, and hence semi-classical injectance is exact at this point which can be seen in Fig. 5.13a. Point F is marked at the same value of kl as in Fig. 5.12. But, in Fig. 5.12, the phase drop at G is now not sharp enough to decrease to π. It is due to tuning the potential, that the subloop has now reduced in area, and consequently the drop has also reduced and will eventually be reduced to a cusp as V is decreased further. At such point $\rho^e(1, E)$ is not equal to $\rho^s(1, E)$, i.e., semi-classical injectance is not exact in spite of a phase drop. This can be seen in Fig. 5.13b where point G is marked at the same value of kl, as in Fig. 5.12. One would have expected that when we decrease the potential and make it weaker, semi-classical behavior will be favored. But on the contrary, for stronger potential the semi-classical injectance is exact at point N in Fig. 5.10b, while for a weaker potential in the same system the semi-classical injectance is not exact at point G in Fig. 5.12. Therefore, the drops in scattering phase shift of mesoscopic system, originating from subloops in A-D involves completely new physics. One has to discard the usual concept of semi-classical regimes wherein the de Broglie wavelength of the electron is much smaller than the scale of the potential, mathematically expressed by Eq. (5.13).

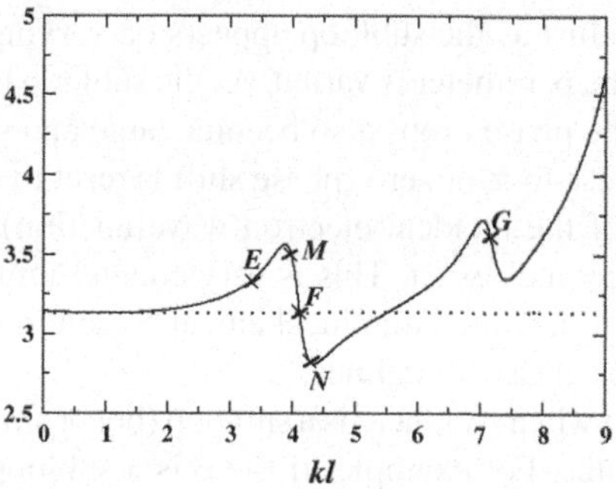

Fig. 5.12 Plot of reflection phase shift $\theta_{r_{11}}$ versus kl for the three prong potential. Here $l = 1$, $l_2 = 5l$, $e = 1$ and $eVl = -1000$

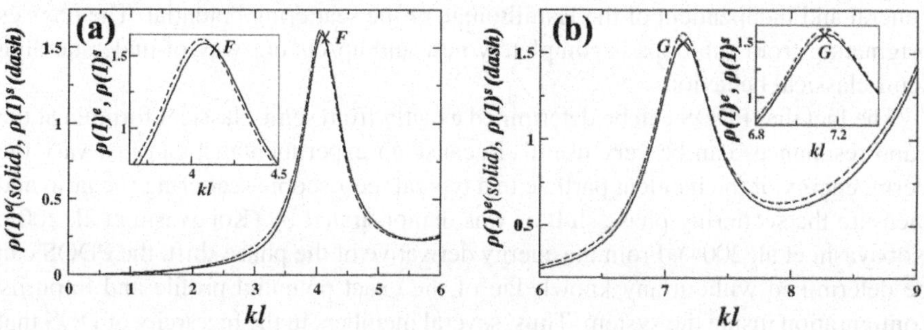

Fig. 5.13 Plot of exact injectance $\rho^e(1, E)$ (solid line) and semi-classical injectance $\rho^s(1, E)$ (dashed line) as a function of kl for the three prong potential. The peaks in the injectance are shown separately, **a** shows the first peak, **b** shows the second peak, for the same parameters as in Fig. 5.12. The insets show the magnified curves at points F and G

5.4 Summary

A-D of scattering matrix elements are drawn for different model potentials and can also be done for experimental data (Satpathy and Singha Deo 2016). Several conclusions can be drawn from the A-Ds without referring to the Hamiltonian or to the scattering potential. In 1D, 2D and 3D, the A-D trajectory encircles the phase singularity. But in mesoscopic systems, we find that the A-D can develop subloops. A subloop does not enclose the phase singularity at the origin and hence topologically allowed. A subloop can therefore appear or disappear on varying some parameter. It does not matter what parameter (say E) is varied to obtain the A-D or what parameter (say V) is varied to make the subloop disappear. Many unexplained features that has puzzled physicists for some time can be explained by the appearance and disappearance of a subloop. When the subloop appears there will be a gradual drop in the scattering phase shift and when the subloop disappear the drop will also disappear. Hence, appearance and disappearance of phase drop is also very natural and poses no conceptual problem. Just as the subloop appears on varying some parameter, it can also grow in size as the parameter is varied. As the subloop becomes large and comes closer to the origin the phase drops also become large and sharp making the scattering phase shift decrease to π or zero (phase shift is reset to initial value independent of the initial phase of the incident electron wavefunction) and then semi-classical injectance or FSR becomes exact. This is very counter intuitive as the strong phase drop signifies onset of pure quantum mechanical behavior, while FSR is expected to become exact in semi-classical regimes.

Also we prove that whenever there is a subloop (big or small) there will be negative partial density of states. For example, if there is a subloop in the range ΔE, then there is also negative partial density of states in the range ΔE. Conclusive evidence of negative partial density of states in a real system has never been reported before. Since all these results are drawn from the properties of the A-D, the results are general and independent of the Hamiltonian or the scattering potential. The physics originating from subloops is completely new and upsets our way of understanding semi-classical behavior.

The fact that PDOS can be determined exactly from semi-classical formula at the Fano resonances can be very useful. Because an experimentalist can just vary the Fermi energy of the incident particle in a typical mesoscopic scattering scenario and measure the scattering phase shift as was demonstrated by (Kobayashi et al. 2003; Kobayashi et al. 2004). From the energy derivative of the phase shift, the PDOS can be determined without any knowledge of the exact potential profile and impurity configuration inside the system. Thus, several members in the hierarchy of DOS that determine thermodynamic and transport properties of the sample can be known. This will be especially useful because mesoscopic systems are not in the ergodic regime allowing us to average over impurity configurations.

Appendix A

From Eq. (5.8), we can write

$$\rho_{pd}(3,1) = -\frac{1}{2\pi}\int_{sample} d\mathbf{r}^3 |t_{31}|^2 \frac{\delta\theta_{t_{31}}}{\delta V(\mathbf{r})} \tag{5.34}$$

In Eq. (5.34), using the well-known semi-classical approximation $\int_{sample} d\mathbf{r}^3 \frac{\delta t_{31}}{\delta V(\mathbf{r})} \cong -\frac{dt_{31}}{dE}$, we get (see Eq. 5.13)

$$\rho_{pd}(3,1) \approx \frac{1}{2\pi}|t_{31}|^2 \frac{d\theta_{t_{31}}}{dE} \tag{5.35}$$

Now, for a contour C' traced when energy is varied from 0 to E_1,

$$\int_{C'} d\theta_{t_{31}} = \int_0^{E_1} \frac{d\theta_{t_{31}}}{dE} dE$$

$$= \int_0^{E_1} \frac{\frac{1}{2\pi}|t_{31}|^2 \frac{d\theta_{t_{31}}}{dE}}{\frac{1}{2\pi}|t_{31}|^2} dE$$

Using Eq. 5.35 $$\int_{C'} d\theta_{t_{31}} \approx 2\pi \int_0^{E_1} \frac{\rho_{pd}(3,1)}{|t_{31}|^2} dE \qquad (5.36)$$

Appendix B

We know that in a scattering problem increasing incident energy by dE is equivalent to decreasing the potential globally by a constant amount $\Delta\varepsilon$, such that $dE = -e\Delta\varepsilon$, where e is particle charge that we will set to 1 to simplify our arguments. That is, the new potential is $V'(\mathbf{r}) = V(\mathbf{r}) - \Delta\varepsilon$. Hence, if we can generate a closed subloop in the A-D by varying E, then we can also do so by globally changing the potential and for such a closed contour like $ABQFA$ in Fig. 5.7c,

$$\oint_{ABQFA} \delta\theta_{s_{\alpha\beta}} = 0$$

i.e., $$-\oint_{ABQFA} \int_{global} \frac{\delta\theta_{s_{\alpha\beta}}}{\delta V(\mathbf{r})} \Delta\varepsilon d\mathbf{r}^3 = 0 \qquad (5.37)$$

Now we replace the global integration over $\mathbf{r}$ by an integration over the sample or the scattering region only, since we have seen that it can be done in case of closed contours or inside an integration of the type $\oint_C$ in Eq. (5.37). This has been discussed in Eq. (5.23).

$$\therefore \quad -\oint_C \int_{sample} \frac{|s_{\alpha\beta}|^2}{|s_{\alpha\beta}|^2} \frac{\delta\theta_{s_{\alpha\beta}}}{\delta V(\mathbf{r})} \Delta\varepsilon d\mathbf{r}^3 = 0$$

or, $$-\oint_C \frac{1}{|s_{\alpha\beta}|^2} \Delta\varepsilon \int_{sample} |s_{\alpha\beta}|^2 \frac{\delta\theta_{s_{\alpha\beta}}}{\delta V(\mathbf{r})} d\mathbf{r}^3 = 0$$

or, $$2\pi \oint_C \frac{\rho_{pd}(\alpha,\beta)}{|s_{\alpha\beta}|^2} \Delta\varepsilon = 0$$ Using Eq. (5.8)

Therefore, the R.H.S of Eq. (5.25) is justified if an electronic charge is multiplied to the numerator.

References

1. M.V. Berry, J. Mod. Opt. **45**, 1845–1858 (1998)
2. M. Büttiker, *Time in Quantum Mechanics* (Springer, 2001) pp. 279–303
3. R. Englman, A. Yahalom, Phys. Rev. B **61**, 2716 (2000)
4. A.M. Jayannavar, Phys. Rev. B **49**, 14718 (1994)
5. K. Kobayashi, H. Aikawa, S. Katsumoto, Y. Iye, Phys. Rev. B **68**, 235–304 (2003)
6. K. Kobayashi, H. Aikawa, A. Sano, S. Katsumoto, Y. Iye, Phys. Rev. B **70**, 035–319 (2004)
7. R. Landauer, T. Martin, Rev. Mod. Phys. **66**, 217–228 (1994)
8. J.F. Nye, M.V. Berry, Proc. Roy. Soc. Lond. A **336**, 165–190 (1974)
9. U. Satpathy, P. Singha Deo, Ann. Phys. **375**, 491 (2016)
10. R. Schuster, E. Buks, M. Heiblum, D. Mahalu, V. Umansky, H. Shtrikman, Nature (London) **385**, 417–420 (1997)
11. P. Singha Deo, Int. J. Mod. Phys. B **19**, 899 (2005)
12. P. Singha Deo, Pramana J. Phys. **58**, 195–203 (2002)
13. P. Singha Deo, Solid State Commun. **107**, 69 (1998)
14. A.L. Yeyati, M. Büttiker, Phys. Rev. B **62**, 7307 (2000)

Chapter 6
Time Travel

Time in quantum mechanics appear as a parameter and there is no self-adjoint time operator consistently defined yet. This is not a serious problem as experimentalists only measure time intervals. This suits mesoscopic physics wherein we have a system which acts as a quantum scatterer coupled to reservoirs via perfectly conducting leads. The reservoirs are completely thermalized and classical and inside the reservoirs time will be just the classically recorded time. The reservoirs also ensure that states do not naturally form linear superpositions in the leads and this results in text book style scattering phenomenon that can be studied by constructing an incident wavepacket or plane partial waves. The electron propagation from one lead to another lead takes place via scattering through the sample and this dynamics is purely guided by quantum mechanics. Thus, a propagation time or a traversal time does not violate any principle. However, theoretical problems remain and that is what we intend to address in this chapter. Quantum mechanics give this measured time intervals very correctly in the semi-classical limit (Landauer and Martin 1994). The low-energy quantum limit is so far not understood in 1D, 2D and 3D since calculated time interval is not always consistent with the Copenhagen interpretation of quantum mechanics. It is known that quantum mechanics starts from a very different set of axioms and does not have to respect theory of relativity within the single-particle coherence length. At the low energies of the mesoscopic systems, Schrodinger (Sc.) equation determines the dynamics within this single particle coherence length and it is completely independent of the speed of light. Beyond this coherence length, all phenomena have to respect special theory of relativity and information speed is limited by the speed of light. We go beyond 1D, 2D and 3D and look into quasi1D (Q1D). Also when we say that negative times are possible then we do not make any reference to theory of relativity. But even within the axioms of quantum mechanics the meaning of negative time has not been completely explained and we want to point out some realistic examples to show that negative times are completely consistent with quantum mechanics.

P. Singha Deo, *Mesoscopic Route to Time Travel*,
https://doi.org/10.1007/978-981-16-4465-8_6

There are two traversal times defined in quantum mechanics that are called Larmor precession time (LPT) and Wigner delay time (WDT). There are many other ways of defining time that correspond to different physical situations, but they cannot be called traversal time because they are not consistent with the number of intermediate states through which a particle is propagated or traversed in time. Our definition of traversal time is that it has to be consistent with the density of states (DOS) up to a constant factor of h (Plank's constant). This definition is consistent with the definition of traversal time for tunneling under the barrier (Buttikker and Landauer 1982) where one considers the time interval in which the tunneling particle interacts with the entire length of the barrier. For tunneling modes, higher order multiple reflections are more strongly damped and so only the length of the barrier matters. We are not considering the situation of tunneling but resonant situations, namely, Fano resonances. Our subject of study is the time interval consistent with DOS and how they are related to the propagation of a wavepacket. Thus, the only two candidates are LPT and WDT. Different candidates for traversal time may have different mathematical forms but should lead to the same quantitative value in order to be consistent. The reason for their different forms is explained below.

In this paragraph, we restrict our discussion to 1D. The derivation of LPT in Chap. 3, starts from a monochromatic plane wave or a stationary state and derives an expression for traversal time. Hence, it starts from the time independent Sc. equation and derives a time. LPT is physically the average traversal time of electrons in the stationary beam and such electrons in a stationary beam cannot be used to send a signal. However, it is completely consistent with DOS at all energies. If not restricted by having to carry a signal or information, even in classical physics one can exceed the speed of light. A signal can be sent by using a wavepacket, for example, a large or a small wavepacket can mean different things. We will show that the LPT is also relevant for signal propagation time. Derivation of WDT actually considers the propagation of a wavepacket in time to derive a traversal time. The LPT is exact but the WDT is approximate using stationary phase approximation. If stationary phase approximation is valid at some energy range then the LPT and the WDT are quantitatively the same, positive definite and fully consistent with DOS. This happens at energies when quantum effects are moderate confirming the Copenhagen interpretation as will be explained below. At low energies, when quantum effects dominate, then LPT and WDT do not give the same result. LPT is exact (as the derivation is exact but explicit calculations at low energies to show its negativity has not been done), while WDT is approximate and can become negative (at these low energies, WDT is no longer consistent with DOS and so its negativity need not mean anything physical, especially as stationary phase approximation is involved). This will be further explained in the next four paragraphs.

We describe below the derivation of WDT as we want to clearly outline the nature of the approximation used. Consider a 1D wavepacket.

$$u(x, \tau) = \frac{1}{\sqrt{2\pi}} \int_{-\infty}^{\infty} a(k) e^{i(kx-\omega\tau)} dk \tag{6.1}$$

Here, $a(k)$ is the weight of the k-component $e^{i(kx-\omega\tau)}$. $a(k)$ can be determined as

$$a(k) = \frac{1}{\sqrt{2\pi}} \int_{-\infty}^{\infty} u(x, 0) e^{-ikx} dx \tag{6.2}$$

For a Gaussian wavepacket,

$$u(x, 0) = e^{-\sigma^2 x^2 + ik_0 x} \tag{6.3}$$

Then

$$a(k) = \frac{1}{\sigma\sqrt{2}} e^{\frac{-(k-k_0)^2}{4\sigma^2}} \tag{6.4}$$

Therefore, substituting from Eq. 6.4 in Eq. 6.1,

$$u(x, \tau) = \frac{1}{\sqrt{2\pi}} \int_{-\infty}^{\infty} dk \frac{1}{\sigma\sqrt{2}} e^{\frac{-(k-k_0)^2}{4\sigma^2}} e^{i(kx-\omega\tau)} \tag{6.5}$$

Therefore,

$$u_{tr}(x + L, \tau + \tau_0 + \Delta\tau) = \frac{1}{\sqrt{2\pi}} \int_{-\infty}^{\infty} dk \frac{1}{\sigma\sqrt{2}} t(k) e^{\frac{-(k-k_0)^2}{4\sigma^2}} e^{ik(x+L)-i\omega(\tau+\tau_0+\Delta\tau)} \tag{6.6}$$

Here, $t(k)$ is transmission amplitude of the length L. τ_0 is the time taken to transmit in the absence of potential in the region of length L i.e., when $t(k) = e^{ikL}$. $\tau_0 + \Delta\tau$ is the time taken to transmit in the presence of potential. Therefore,

$$u_{tr}(x + L, \tau + \tau_0 + \Delta\tau) =$$

$$\frac{1}{\sqrt{2\pi}} \int_{-\infty}^{\infty} dk \frac{1}{\sigma\sqrt{2}} |t(k)| e^{i\eta_k} e^{\frac{-(k-k_0)^2}{4\sigma^2}} e^{ik(x+L)-i\omega(\tau+\tau_0+\Delta\tau)} \tag{6.7}$$

where $t(k) = |t(k)|e^{i\eta_k}$. Now if σ^2 is very small then $|t(k)|$ can be taken to be independent of k and $|t(k)| = t$. It implies that the transmission modulus does not disperse the wavepacket, but the transmission phase shift can still disperse it. From Eq. 6.7,

$$u_{tr}(x + L, \tau + \tau_0 + \Delta\tau) = \frac{1}{2\sqrt{\pi}} \frac{1}{\sigma} t \int_{-\infty}^{\infty} dk e^{\frac{-(k-k_0)^2}{4\sigma^2}} e^{i\eta_k + ik(x+L) - i\omega(\tau+\tau_0+\Delta\tau)} \tag{6.8}$$

So for the wavepacket to remain undispersed, the weight of a particular component must remain unchanged implying

$$kx - \omega\tau = \eta_k + k(x + L) - \omega(\tau + \tau_0 + \Delta\tau) = \text{constant} \tag{6.9}$$

$$\text{or,}\ \eta_k + kL - \omega(\tau_0 + \Delta\tau) = 0$$

$$\text{or,}\ \frac{d\eta_k}{d\omega} + L\frac{dk}{d\omega} = \tau_0 + \Delta\tau \tag{6.10}$$

Also, in the absence of scatterer

$$\therefore\ kx - \omega\tau = k(x + L) - \omega(\tau + \tau_0)$$

$$\text{or,}\ kL - \omega\tau_0 = 0$$

$$\text{or,}\ L\frac{dk}{d\omega} = \tau_0 \tag{6.11}$$

Therefore, from Eq. 6.10,

$$\frac{d\eta_k}{d\omega} = \Delta\tau$$

$$\frac{d\eta_k}{dE}\frac{dE}{d\omega} = \Delta\tau$$

$$\hbar\frac{d\eta_k}{dE} = \Delta\tau^W \qquad \text{as}\ E = \hbar\omega \tag{6.12}$$

$\Delta\tau^W$ is WDT.

If dispersion of wavepacket becomes stronger in the low-energy quantum regime, then to preserve the approximations used in Eqs. 6.8 and 6.9, we have to take a smaller momentum range to make the wavepacket and this may not be possible as we go up to the low energies where $\Delta\tau$ may become negative in 1D. So, a clear meaning of negative WDT has not emerged till date. All wavepackets in 1D, 2D and 3D, disperse and so Eq. 6.9 correspond to an approximation called stationary phase approximation. Hence, in a quantum regime if stationary phase approximation fails, then Eq. 6.12 will not give the traversal time, while at higher energies in the semi-classical limit (de Broglie wavelength is smaller than the spatial scale in which the potential varies), it correctly gives the traversal time. In fact, the reverse can also be proved. That is iff Eq. 6.12 gives the correct traversal time then stationary phase approximation is valid and the wavepacket will propagate undispersed. The treatment can be naturally extended to higher dimensional S-matrices.

Consider the three prong potential shown in Fig. 5.6 and explained in details in the figure caption. $k = \sqrt{\frac{2mE}{\hbar^2}}$ and $q = \sqrt{\frac{2m(E-V)}{\hbar^2}}$ are wavevectors along the thin lines and the thick solid lines, respectively. An incident wavepacket along lead 1 will split into three wavepackets that are reflected and transmitted to leads 2 and 3. The WDT $\Delta\tau_{31}^W$ for the wavepacket coming out of lead 3 will be

$$\hbar\frac{d\theta_{t_{31}}}{dE} = \Delta\tau_{31}^W \tag{6.13}$$

Here, $\theta_{t_{31}}$ is the scattering phase shift of the electrons transmitted from 1 to 3. Note that the propagation of a wavepacket in time has been used to derive Eq. 6.13. An axiom in quantum mechanics states that if we project one particle or a small number of particles into a scatterer, then this particle may get reflected or transmitted to 2 or to 3 in a random manner. But if we repeat the process with a statistically large number of electrons, then it will produce the interference pattern and all observable quantities like r_{11}, t_{21} and t_{31} can be obtained from the stationary state solutions of the time-independent Sc. equation. In other words, η_k in Eq. 6.12 or $\theta_{t_{31}}$ in Eq. 6.13 is in practice always obtained from the time-independent Sc. equation.

In Chap. 3, we start from the time-independent Sc. equation and derive a traversal time called LPT as (follows from simplifying Eq. 3.8)

$$\Delta\tau_{lpt}(3, 1, E) = -\hbar \int_{sample} d\vec{r} \frac{\delta\theta_{t_{31}}}{\delta V(\vec{r})} \tag{6.14}$$

for particles transmitted from 1 to 3. $|t_{31}|^2$ fraction of the incident electrons get transmitted like this. Here, $\frac{\delta}{\delta V(\vec{r})}$ means a functional derivative with respect to the local potential $V(\vec{r})$. $\int_{sample} d\vec{r}$ means an integration over all the coordinates of the sample or the scattering region where the wavefunction deviates from the typical asymptotic wavefunctions of a scattering problem. This effectively means that we change $V(\vec{r})$ at all points inside the sample by an infinitesimal amount of $e\Delta\epsilon$ where e is the incident particle charge. In case of Fig. 5.6, the sample or the scattering region is simply the region shown in thick lines. The derivation of Eq. 6.14, unlike the derivation of WDT in Eqs. 6.12 or 6.13 does not use any approximation. So when WDT is correct, it has to give the same physical quantity as LPT. So if a signal can be sent in WDT (the correct one), it can be also sent in LPT.

At this point, we would like to clearly define the three terms we use subsequently, in this chapter. First is the "traversal time" which will be used in a literary sense that it is a time associated with a particle going from one point to another point traversing the intermediate states connecting the two points. So, for a time interval to qualify as traversal time it has to be related to DOS. Second is the "LPT" which is exact and always sums up to give the correct DOS. Third is the "WDT" which is correct if stationary phase approximation is correct and not otherwise. However, from now on, we will not use terms like "correct WDT" and "incorrect WDT". Correctness or incorrectness will be born by its comparison with LPT. If WDT is correct then it has to be equal to LPT.

We will first show that in 1D at low energy, LPT remain positive, while WDT may become negative. We use Burgers circuit (BC) to prove this. Next, we will show that in Q1D, one can find frequently occurring regimes where LPT and WDT can both be identically negative. That is we will prove that in Q1D there are regimes wherein

$$\Delta\tau_{lpt}(3, 1, E) = \Delta\tau_{31}^{W} < 0 \tag{6.15}$$

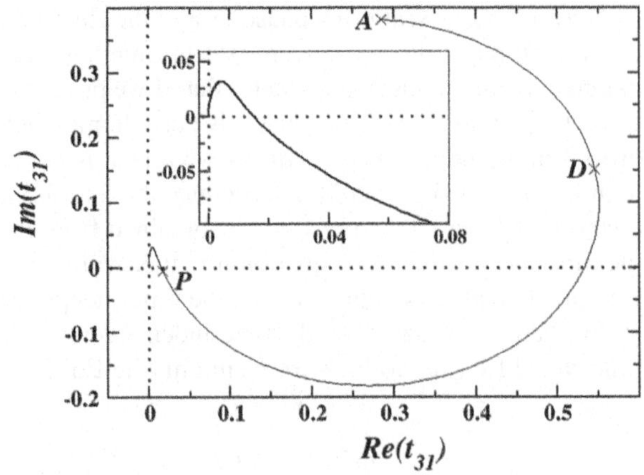

Fig. 6.1 Argand diagram for transmission amplitude t_{31} for the three prong potential shown in Fig. 5.6. The A-D is obtained by varying the wavevector kl from 0 to 5. Here, $l_1 = l_3 = l$, $l_2 = 5l$ and $eVl = -1000$. The inset show a magnified picture of the region near the origin

Before showing this, we would like to summarize why such a demonstration is interesting. First of all, it is new and no such previous examples or a system that exhibit this can be cited. Second, it implies that $\Delta\tau_{lpt}(3, 1, E)$ can be negative and such a quantitative negative value for it has not been obtained before. $\Delta\tau_{lpt}(3, 1, E) < 0$ does not prove anything with respect to signal propagation time unless one can show $\Delta\tau_{lpt}(3, 1, E) = \Delta\tau_{31}{}^{W} < 0$ because only then one can claim that a wavepacket or some change in spatial probability distribution of electrons can be transmitted to negative times within the single particle coherence length.

Since quantum mechanics and relativity are anyway not consistent with each other but both very successful in their own regimes, it has been always believed that superluminal times are possible in quantum mechanics. But a concrete demonstration consistent with quantum mechanics (no need to bring relativity in the picture and expressed in Eq. 6.15) itself has not been shown so far. We will use the three prong potential to illustrate our results. The proof is general, and can be applied to any scattering matrix element $S_{\alpha\beta}$ that make loops in the first Riemann surface and in mesoscopic systems one can find a large class of systems that exhibit this essentially due to Fano resonance. BC uses the analyticity of the complex scattering matrix elements and is therefore more general than quantum mechanics and so quantum mechanics has to respect it.

In Fig. 6.1, we give a part of the Argand diagram (A-D) in Fig. 5.7c which is the A-D for t_{31} and, in Fig. 6.2 similarly, we give a part of Fig. 5.7d which is a plot of $\theta_{t_{31}}$ as a function of the wavevector for the same range as in Fig. 6.1. Note that in Fig. 6.2 around the point marked P, the scattering phase shift decreases with kl that is $\frac{d\theta_{t_{31}}}{dE}$ $(= \frac{ml}{\hbar^2 k}\frac{d\theta_{t_{31}}}{d(kl)})$ will be negative. The corresponding point in the A-D is also marked P in Fig. 6.1. One can calculate $\theta_{t_{31}} = Arctan[\frac{Im(t_{31})}{Re(t_{31})}]$ at point P in Fig. 6.1 at close by points to see from the A-D as well that $\theta_{t_{31}}$ has a negative slope here. This negative slope does not have anything to do with the three prong potential and it can be observed for all values of l_2 at the same energy and of similar magnitude. For example, by making $l_2 \to 0$ in Fig. 5.6, one will get the 1D limit

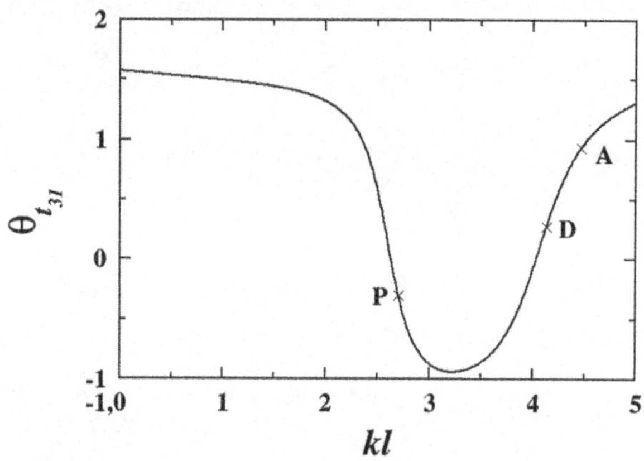

Fig. 6.2 Plot of transmission phase shift $\theta_{t_{31}}$ corresponding to transmission amplitude t_{31} for the three prong potential shown in Fig. 5.6, as a function of kl. Here, $l_1 = l_3 = l, l_2 = 5l$ and $eVl = -1000$

where too one can check and find this negative slope at the same energy and of similar magnitude. This negative slope in 1D has been the matter of study in the past with respect to traversal time, and there is no conclusive understanding as to whether it means something physical or is an artifact of the stationary phase approximation (Landauer 1994). It can be seen in a 1D potential like a square barrier or a square well potential ($l_2 \to 0$ limit of Fig. 5.6) where it happens only once as the energy or the momentum is increased from 0. We can understand the origin of this negative slope from the A-D using BC. At zero energy the transmission is zero even for an infinitesimal potential in the scattering region, implying $Im(t_{31})$ and $Re(t_{31})$ are both zero. This is a point where $\theta_{t_{31}}$ is singular. So the A-D starts or originates from the singularity and comes out of the singularity as the incident energy increases from 0. Now since the absolute value of t_{31} is bounded, the A-D has to curl around the singularity either in clockwise direction or anticlockwise direction. Clockwise will mean phase will decrease whereas anticlockwise will mean phase will increase with energy as suggested by Eq. 5.1. Initially, however, when the A-D trajectory emerges from the singularity, it is neither clockwise nor anticlockwise with respect to it. After a while, it turns around and before becoming anticlockwise with respect to the origin it goes through a small region around point P where it is clockwise, and this is what creates the negative slope. Once it has gone anticlockwise it never turns back again in 1D. And so after this, initial drop phase monotonously increases. The phase of the wavefunction increases implies rotation of the wavefunction in Hilbert space unidirectionally which is like a notion of time increasing monotonously and positively, as time evolution is guided by a factor $e^{iEt/\hbar}$. Now in case of the three prong potential, we would like to make the following observations that for some reason has not been observed in 1D essentially because making these observations is not very revealing in 1D. Our observations for the three prong potential will become more relevant as the chapter progresses.

First of all, one can keep the energy (or kl) fixed at the value corresponding to the point P in Fig. 6.1, and decrease (or increase) the potential globally by constant amounts to generate identically same A-D starting from P. Energy increasing cor-

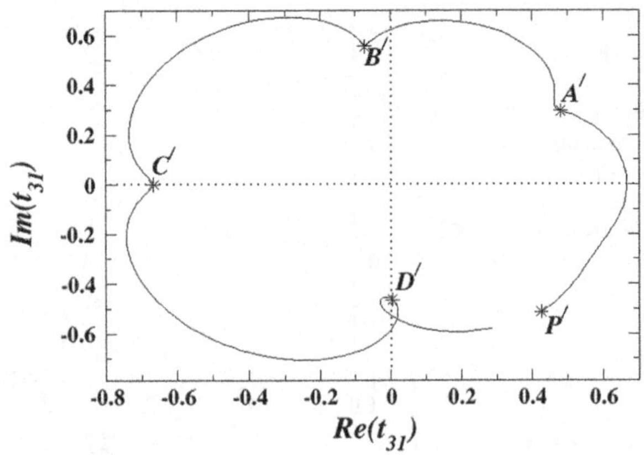

Fig. 6.3 Argand diagram for transmission amplitude t_{31} for the three prong potential shown in Fig. 5.6. The A-D is obtained by varying the potential eVl from 0.0 to -25. It starts from the point marked $P^{/}$. Here, $l_1 = l_3 = l$, $l_2 = 5l$ and $kl = 2.7$ which are the parameters corresponding to the point P in Figs. 6.1 and 6.2. Even when kl value is taken close to zero rather than 2.7, the point $P^{/}$ does not get close to the origin. The phase singularity is at the origin where $t_{31} = 0$

responds to potential decreasing globally and vice versa. But this is a very obvious observation. Now to come to the point, suppose we keep the energy fixed at that corresponding to point P and decrease the potential only in the thick region in Fig. 5.6 (that is only in the scatterer) by constant amounts starting from 0. Unlike the A-D of Fig. 6.1, this A-D shown in Fig. 6.3 will not start from the singularity but start from a point completely different from P. Because when the potential is zero, even a mode with infinitesimal energy or momentum gets fully transmitted. So even if the point P in Fig. 6.1 is taken arbitrarily close to the origin, the A-D obtained by varying the potential V (in the thick region of Fig. 5.6, starting from $V = 0$) will not start from the origin. It starts from P$^{/}$ and right from the very start it goes anticlockwise with respect to the origin or with respect to the singularity, its trajectory being completely unaffected by the presence of the singularity. Only when a trajectory emerges from the origin, it can neither be clockwise or anticlockwise with respect to the origin. The scattering phase shift shown in Fig. 6.4 corresponding to this A-D shown in Fig. 6.3, therefore starts increasing from -1 and there is no negative slope near the origin. Only at a higher value of the potential V, i.e., at the point D$^{/}$, it shows a negative slope due to the small subloop at $D^{/}$ in Fig. 6.3. Decreasing the constant potential V only in the thick region of Fig. 5.6 in small steps of $e\Delta\epsilon$ allows us to evaluate the RHS of Eq. 6.14 and thus $\Delta\tau_{lpt}(3, 1, E)$. Here, e is the particle charge that is set to 1 without any loss of generality. Thus, A-D explains why LPT will not be negative at P and can be verified from Eq. 6.14. So at low energies or in the quantum regime, global constant shift in potential and constant shift of potential in the scattering region produce completely different results. Adding up all possible LPT for all the channels like t_{21}, r_{11}, etc., and multiplying with $\frac{1}{h}$, one can get density of states (DOS) that can be independently determined from the internal wavefunction also for

Fig. 6.4 Plot of transmission phase shift $\theta_{t_{31}}$ corresponding to the A-D in Fig. 6.3, as a function of $-eVl$ (V is negative and decreasing)

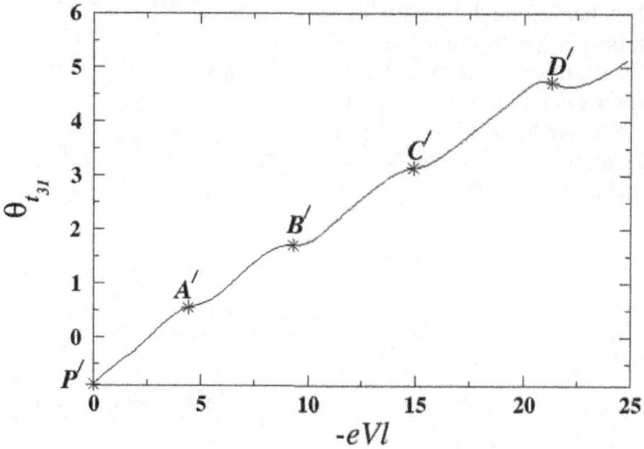

Fig. 6.5 Here, $l_1 = l_3 = l$, $l_2 = 5l$ and $eVl = -1000$. Around three broad minima at $kl = 2.5$, 6.7 and 11.4, Eq. 6.15 is satisfied

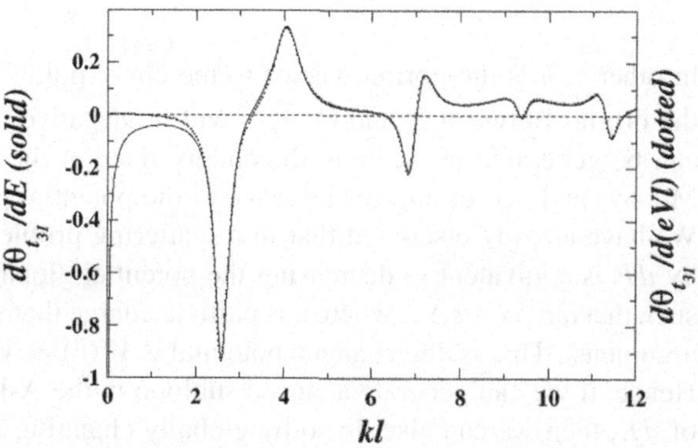

comparison. We have checked that there is perfect agreement of this LPT calculated from Eq. 6.14 with the DOS determined from internal wavefunction. In Fig. 6.5, we plot $\frac{d\theta_{t_{31}}(E)}{dE}$ and $\frac{d\theta_{t_{31}}(E)}{d(eV)}$, where V is the constant potential in the sample region, i.e., the region shown in Fig. 5.6 in bold lines. Both are plotted as a function of kl. It shows $\frac{d\theta_{t_{31}}}{dE}$ is strongly negative near $k = 0$ but $\frac{d\theta_{t_{31}}}{deV}$ is not. This negativity of $\frac{d\theta_{t_{31}}}{dE}$ is due to the failure of stationary phase approximation as it does not agree with LPT. LPT starts from zero and starts going negative from the very start (note that both the derivatives in Fig. 6.5 are calculated at a large fixed value for potential V in the sample). However, LPT does not agree with WDT Eq. 6.15 is not satisfied. However, at three broad minima shown in Fig. 6.5 and mentioned in the figure caption, Eq. 6.15 is satisfied. There are other regions also where $\frac{d\theta_{t_{31}}}{dE}$ and $\frac{d\theta_{t_{31}}}{deV}$ are identical, but since they are not negative, there Eq. 6.15 is not satisfied.

Now in case of Fig. 6.1, if we increase the energy further then the A-D show subloops within a particular Riemann surface and demonstrated in Fig. 6.6. At each subloop, we get a portion of A-D trajectory that is clockwise with respect to the origin. For example, we have highlighted one such subloop in Fig. 6.6 by dotted lines and marked the subloop as MNOM. Between N and O, the trajectory is clockwise or

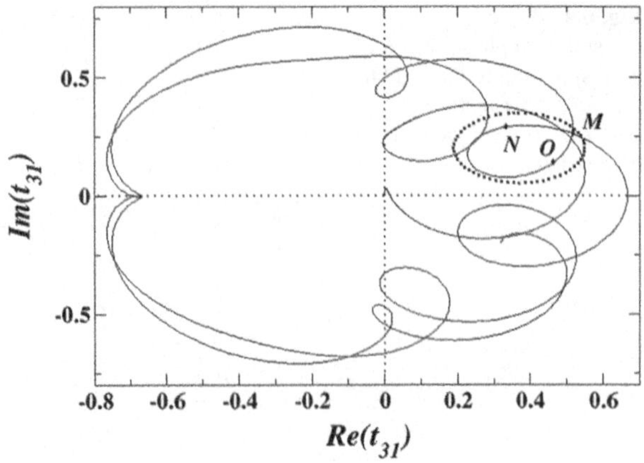

Fig. 6.6 Argand diagram for transmission amplitude t_{31} for the three prong potential shown in Fig. 5.6. kl is varied from 0 to 20, to get this A-D. Here, $l_1 = l_3 = l, l_2 = 5l$ and $eVl = -1000$

in other words the normal vector to the curve points towards the origin and scans the origin. Between N and O, $\frac{d\theta_{t_{31}}}{dE}$ will be negative. The identical A-D of Fig. 6.6 can be generated by keeping the energy fixed at the value corresponding to point M (say) and decreasing (or increasing) the potential globally by constant amounts. We have already discussed that in a scattering problem increasing incident energy by dE is equivalent to decreasing the potential globally by a constant amount $\Delta\varepsilon$, such that $dE = -e\Delta\varepsilon$, where e is particle charge that we will set to 1 to simplify our arguments. That is, the changed potential is $V'(\vec{r}) = V(\vec{r}) - \Delta\varepsilon$ for $-\infty < \vec{r} < \infty$. Hence, if we can generate a closed subloop in the A-D by varying E in small steps of dE, then we can also do so by globally changing the potential in small steps of $\Delta\epsilon$. Decreasing the potential globally is equivalent to increasing the energy and so it will generate the A-D further away from the origin (starting from M and towards N) and increasing the potential is equivalent to decreasing the energy and so it will generate the A-D from M down to the origin. So the subloops can also be generated by decreasing the potential globally (that includes the thick regions as well as in the leads in Fig. 5.6) starting from the V value used in Fig. 6.6 and keeping the energy fixed at the value corresponding to point M in Fig. 6.6. For a closed contour A corresponding to a particular subloop (say MNOM) in Fig. 6.6,

$$\oint_A \Delta\theta_{s_{\alpha\beta}} = 0$$

$$\oint_A \frac{\Delta\theta_{s_{\alpha\beta}}}{\Delta E} dE = 0$$

$$\text{i.e.,} \quad \oint_A \frac{\Delta\theta_{s_{\alpha\beta}}}{\Delta E} dE = -\oint_A \Delta\varepsilon \int_{global} \frac{\delta\theta_{s_{\alpha\beta}}}{\delta V(\vec{r})} d\vec{r} = 0 \tag{6.16}$$

Here, $\int_{global} d\vec{r}$ correspond to an integration over all spatial coordinates. These subloops are consistent with BC and hence the topological structure of the Riemann surface in which the A-D is drawn. Subloops also belong to a certain topological class such that a particular subloop is very fundamental and cannot be removed by varying any parameter (note that, in previous chapter, we said that the presence or absence of a subloop does not have any effect on the conserved quantity I in Eqn. 5.1 and so subloops can occur without violating any principle. Now we are using topological argument to say that once we identify a subloop, it cannot be removed by adjusting any parameter, where of course we are excluding possibilities of changing the system itself like jumping from a closed to an open system. A subloop can grow or reduce in size continuously and can even become infinitesimal when it appears like a cusp). Essentially, it means a multiply connected curve cannot be continuously deformed to a simply connected curve by tuning terms and parameters in the Hamiltonian (there are no phase transitions in small quantum systems and in the regime of single particle coherence length we are anyway not concerned with them). Physically, these subloops originate from Fano resonance and a resonant state cannot be removed by varying any parameter. A resonant state can be populated or de-populated but the total number of states in a system remain conserved. So if we do not increase the incident energy (or decrease the potential globally) but decrease the potential only in the thick region of Fig. 5.6, then we will generate identical number of subloops. The nature of the different subloops may change but the total number of subloops will be the same. This is shown in Fig. 6.7 up to very high values of the potential in the sample region (such high energies are not shown in Fig. 6.6). Therefore, for a closed contour A in Fig. 6.6, one can find the corresponding closed contour B in Fig. 6.7 (essentially one has to find the subloop B coming from a Fano resonance with the same number of nodes as the Fano resonance making subloop A), such that

$$-\oint_B \Delta\varepsilon \int_{sample} \frac{\delta\theta_{s_{\alpha\beta}}}{\delta V(\vec{r})} d\vec{r} = 0 \tag{6.17}$$

By the argument that if a subloop A can be generated by varying the incident energy E, then a corresponding subloop B can be generated by varying the potential uniformly in the sample region in small steps of $e\Delta\epsilon$ and all subloops can be thus paired up.

This only proves that the integrations are equal, i.e.,.

$$\oint_A \frac{\Delta\theta_{s_{\alpha\beta}}}{\Delta E} dE = -\oint_B \Delta\varepsilon \int_{sample} \frac{\delta\theta_{s_{\alpha\beta}}}{\delta V(\vec{r})} d\vec{r} = 0 \tag{6.18}$$

This also proves that the integrands are negative in part of the range in which the relevant parameter is varied to get the closed contours A and B and positive in the rest of it. Thus, this proves that LPT which is always correct, can be negative. All the subloops A in Fig. 6.6 and subloops B in Fig. 6.7 satisfy Eq. 6.18. But this does not explain that around the three minima in Fig. 6.5 at negative values why Eq. 6.15 should be valid. To explain that we have to show that the integrands in Eq. 6.16 can be equal in some regimes. If we can find some regime where the value of the integration becomes independent of the shape and size of the contour A as well as B, then the integrands have to be identical. And also unlike the subloops in Fig. 6.6 and

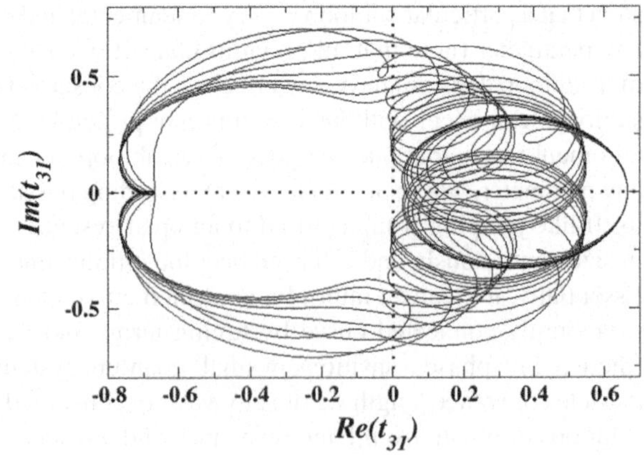

Fig. 6.7 Argand diagram for transmission amplitude t_{31} for the three prong potential shown in Fig. 5.6. The A-D is obtained by varying the potential eVl from -1 to -1000. Here, $l_1 = l_3 = l$, $l_2 = 5l$ and $kl = 4$

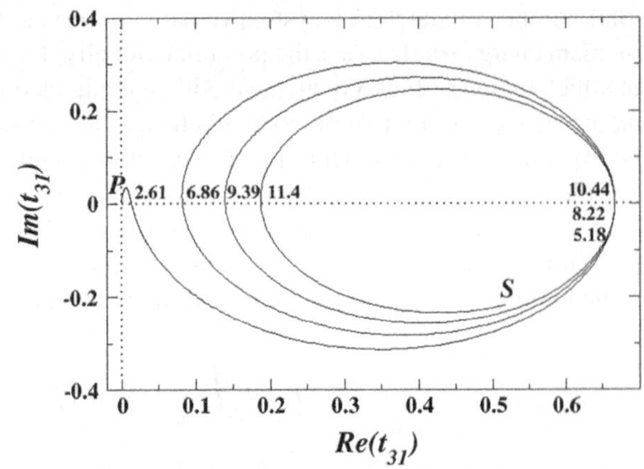

Fig. 6.8 A-D for t_{31} for the system in Fig. 5.6 for the same parameter values as that in Figs. 6.6, and 6.7 with $l_1 = l_3 = 0$. kl is varied from 0 to 12.0 and some kl values like 2.61, 5.18, 6.86, etc., are marked at corresponding points in the A-D. It means the first time the A-D crosses the real axis is at $kl = 2.61$. Then it crosses again at 5.18, then at 6.86 and so on

Fig. 6.7 the trajectory making subloops A and B should not have any discontinuities but should come back onto itself very smoothly or else the integration limits (like the point M in Fig. 6.6 in the trajectory M to N is not identical to the point M in the trajectory O to M) are not identical in all respect. If closed contour A satisfy these two properties then the pairing closed contour B too has to satisfy them. Because the LHS integration cannot be 0 independent of the shape and size of the contour A, while the RHS integration does depend on the shape and size of the contour B. Such smooth subloops A are shown in Fig. 6.8 and smooth subloops B are shown in Fig. 6.9. Relevant parameters are mentioned in the figure caption. Note that, in Fig. 6.9, we have fixed the energy such that $kl = 8.22$ and the initial value of $eVl = -1000$ which implies the trajectory will start exactly from the point marked x in Fig. 6.9, which is identical to the point at $kl = 8.22$ in Fig. 6.8.

In Fig. 6.10, we plot $\frac{d\theta_{t_{31}}(E)}{dE}$ and $\frac{d\theta_{t_{31}}(E)}{deV}$ as a function of kl and above a value of $kl = 8.22$ the two curves become identical as expected from Figs. 6.8 and 6.9. There are broad energy ranges occurring periodically where both curves become negative

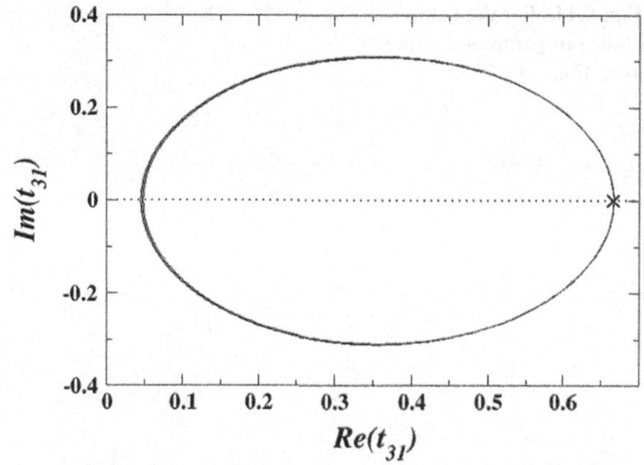

Fig. 6.9 A-D for t_{31} for the system in Fig. 5.6 for the same parameter values as that in Figs. 6.6 and 6.7 with $l_1 = l_3 = 0$, and $kl = 8.22$. eVl is varied from -1000 to -1050. Note that the starting point is marked with a x which is the same as the point marked 8.22 in Fig. 6.8 as the parameter values corresponding to it are the same

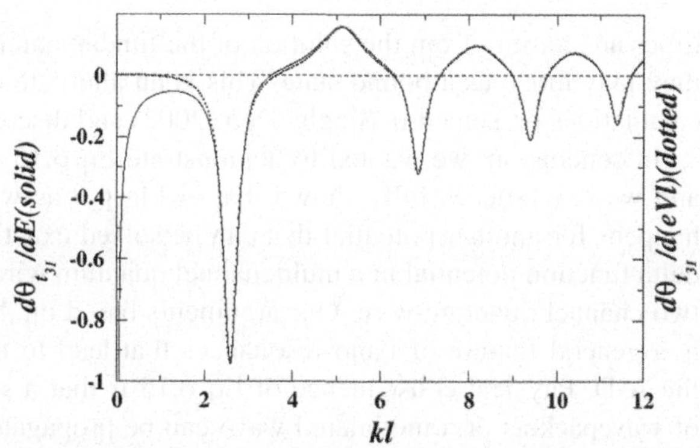

Fig. 6.10 For the same choice of parameter values as in Figs. 6.8 and 6.9

and so Eq. 6.15 is satisfied and this happens up to very high energies. Obviously, if we construct a wavepacket with the modes in these ranges, then that wavepacket will remain undispersed perfectly satisfying the stationary phase approximation. Note that $\theta_{t_{31}}$ (denoted as η_k in Eq. 6.8), plotted in Fig. 6.11 for the parameters corresponding to Fig. 6.10, tends to become a sinusoidal curve at higher values of kl which means $\frac{d\theta_{t_{31}}}{d\omega}$ or $\Delta\tau^W$ (see Eq. 6.8 or 6.9) become a cosine curve. Thus, there is a constant phase difference between the two. In contrast to that, in case of a 1D barrier or well the scattering phase shift increases monotonously, while its derivative oscillates. In case of Fig. 6.5, by the time we reach $kl = 6.5$, the dispersion due to scattering by the two arms of length l in Fig. 5.6, become negligible and we get agreement between $\frac{d\theta_{t_{31}}(E)}{dE}$ and $\frac{d\theta_{t_{31}}(E)}{deV}$. Although they become mostly positive, they do show broad energy ranges where Eq. 6.15 is satisfied.

Electrons in a wavepacket with negative traversal time can move in negative time and so they will behave like positrons. A positron will attract an electron and annihilate each other to form a new particle or quasi particle. High-energy phenomenon of producing photons will not occur in quantum wires as it happens for real positrons. However, this new quasi particle can be quite stable due to the fact that negative

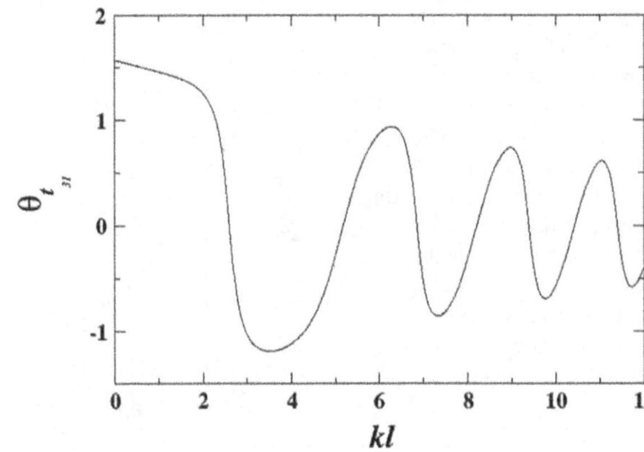

Fig. 6.11 For the same choice of parameter values as in Fig. 6.8

times are coming from the solution of the fundamental equations of motion, and so they may move as a bound state. This is an alternate explanation to the numerical calculations presented in Singha Deo (2002) and discussed in Chap. 5.

In conclusion, we wanted to demonstrate Eq. 6.15 for certain range of energies and we have successfully shown that in Fig. 6.5 as well as in Fig. 6.10. This also happens for another potential that can be solved exactly in Q1D, that of a negative delta function potential in a multichannel quantum wire. We have checked this for a two channel quantum wire. Our arguments based on A-D and BC suggest that this is a general feature of Fano resonances that lead to the formation of subloops in the A-D. Physical consequence of Eq. 6.15 is that a signal in the form of a series of wavepackets or a modulated wave can be propagated in negative time. We also conjecture that one can have a bound state of two electrons as a consequence. This chapter is based on our published work (Singha Deo and Satpathy 2019).

6.1 Reflections

While time travel has been a matter of speculation and contradiction leading to many philosophical questions, we have here by provided a definite mechanism that may carry an information or signal to the past. Such a mechanism was not known so far. One may see Satpathi et al. (2016) for comparison with experimental data. We have therein also pointed out some further confirming experiments can be done. There is no contradiction in accepting the fact that a signal can be sent to the past quantum mechanically and has been talked about since the time of Wigner. We have shown that one can construct a Wigner-type wavepacket that can carry a signal to the past if we restrict to the regime of negative phase drops due to Fano resonances. What could be the consequence of such a signal on our life is a matter of immense speculation that even Hollywood fantasizes on. Any philosophical criticism can always be countered by an equally mind boggling philosophy and so in physics (as a discipline) we restrict ourselves to the nuts and bolts of the mechanism involved and this pays off. We did observe the discipline of polishing our shoes before we went to school as kids although within some "-ism" one can always say that going to school barefooted can do us a lot of good.

Einstein's special theory of relativity (STR) is probably a theory to which there are no known violations as long as we do not want to talk about any form of acceleration. All other theories work very well in some regime and fail beyond that and also has theoretical inconsistencies. This is different from STR or quantum mechanics where so far there are no known inconsistencies. While the standard model or general theory of relativity talks about an incomplete picture, within the low-energy regime as long as single particle wavefunction modulus gives probability (gauge symmetry), quantum mechanics gives a complete picture and within boosted reference frames (Lorentz's symmetry) STR gives a complete picture. Special theory of relativity reveals the fact that past, present and future are all equally real and simultaneity is relative. By moving faster and faster, we go slower to the future, but as we cannot exceed the speed of light, we cannot go to the past. Now moving faster in STR is not a physical process achieved through acceleration and in the twin paradox, return to earth has to be without involving deceleration in STR. Brian Cox and Jeff Forshaw in their book "Why does $E = mc^2$?" has done the calculation for such a physical process involving acceleration and deceleration using general theory of relativity (GTR) and the age difference between the twins is way off from what we get from STR. STR reveals just the fact that nature has a symmetry captured by a transformation called boost which is at the heart of this age difference. Now physical acceleration is not very well understood yet and it is not simultaneously consistent with GTR and quantum mechanics. Long ago, the Greek philosopher Zeno had argued in terms of deceleration that motion is undefined. He considered a process in which in subsequent time intervals a traveller always goes half of his previous step. Today we analyze his setup using calculus and we understand the problem in terms of converging series but that does not mean we have solved the problem of acceleration and defined motion. Today we argue in terms of Unruh effect and Killing vectors and one can emphatically say that motion is undefined. In fact, it is related to our poor understanding of the continuum wherein we do not know if there is an answer to the following question. If a bird physically flies from point A to point B, we can magnify and start seeing its intermediate steps but is there a limit to how many intermediate steps exist? If there is a limit or a finite bound then the bird is doing some vanishing acts at certain points. Symmetry just captures the geometry of the manifold in which the bird flies and does not care for vanishing acts or continuous motion. This however does not mean that the symmetry captured by boost is to be questioned. That so far is beyond all doubts.

Same can be said about quantum mechanics (up to first quantization). It reveals that nature has a symmetry and that particles like waves has a gauge symmetry. Physical processes has to be understood with reference to this symmetry but there are plenty of physical phenomenon (even at low energies) involving transient phenomenon, bistabilities and shocks that are beyond the axioms of quantum mechanics. But that basic symmetry has never been doubted.

The only contradiction we find in quantum mechanics is when we say that the Universe has a Hamiltonian (well then there should be no such thing as a shock wave in this universe but that would be over simplifying the arguments). So when particles are incident on a 3D scatterer, we formulate in terms of a plane wave that satisfies some Sc. equation. We have argued in Chap. 1 and is also well known that it does not. Then we say that this plane wave can be expanded into partial waves and thus we get a LHS and the RHS of an equation wherein the RHS satisfy a Sc. equation term by term, but the LHS does not. However that is justified by deriving the optical theorem and scattering cross section that fits with experiments. On the other hand, mesoscopic physics reveals the following peculiarity. We can start with them as independent waves that do not form linear superpositions. We do not have to sum them up and equate the sum to something. Inside the scatterer it develops states that have nothing to do with the states of the sample Hamiltonian (obvious statement). Now it is possible to think of an effective Hamiltonian of the sample by correcting the sample Hamiltonian with some self-energy terms and it is this effective Hamiltonian that determines the states inside the scatterer, but even this is not fully defined below a certain energy (which in case of Fig. 6.5 is $kl = 6.5$ and for Fig. 6.10 is $kl = 8.22$). On the other hand, at all energies (energies admissible in quantum mechanics as we know mesoscopic systems are built with semiconductors and metals and electron energies cannot exceed the work function of these materials) the states are locally described completely by the functional derivative of the local potentials. The potential landscape or potential scape therefore offers a very good manifold in which quantum mechanics is well defined and a manifold picture is something that quantum mechanics needs very badly if ever it has to become consistent with classical theories at a geometrical level. Of course, there are regimes wherein the self-energy due to coupling to leads become negligibly small and then we get a closed system described by a Hamiltonian. But if we extend that argument by saying the Universe is a closed system described by a Hamiltonian then we again land with the problem that plane waves incident from outside has to follow from some Hamiltonian dynamics. Mesoscopic physics has shown us that this is redundant. However, we leave it as a reflection and much more thought as well as work is needed in this direction.

References

1. M. Buttikker, R. Landauer, Phys. Rev. Lett. **49**, 1739 (1982)
2. R. Landauer, Th. Martin, Rev. Mod. Phys. **66**, 217–228 (1994)
3. P. Singha Deo, Pramana J. Phys. **58**, 195–203 (2002)
4. P. Singha Deo, U. Satpathy, Results Phys. **12**, 1506 (2019)

The manufacturer's authorised representative in the EU is Springer Nature Customer Service Centre GmbH, Europaplatz 3, 69115 Heidelberg, Germany. If you have any concerns regarding our products, please contact ProductSafety@springernature.com

Printed and bound by CPI Group (UK) Ltd, Croydon, CR0 4YY
30/09/2026
02225309-0002